Service Excellence for Sustainability

Nur Fazidah Elias · Ruzzakiah Jenal ·
Hazilah Mohd Amin · Hazura Mohamed ·
Siti Aishah Hanawi
Editors

Service Excellence for Sustainability

Lessons from Malaysia, Japan, and Taiwan

Springer

Editors
Nur Fazidah Elias
Information Science and Technology
Universiti Kebangsaan Malaysia
Bangi, Selangor, Malaysia

Ruzzakiah Jenal
Information Science and Technology
Universiti Kebangsaan Malaysia
Bangi, Selangor, Malaysia

Hazilah Mohd Amin
Information Science and Technology
Universiti Kebangsaan Malaysia
Bangi, Selangor, Malaysia

Hazura Mohamed
Information Science and Technology
Universiti Kebangsaan Malaysia
Bangi, Selangor, Malaysia

Siti Aishah Hanawi
Information Science and Technology
Universiti Kebangsaan Malaysia
Bangi, Selangor, Malaysia

ISBN 978-981-16-2578-7 ISBN 978-981-16-2579-4 (eBook)
https://doi.org/10.1007/978-981-16-2579-4

This Springer imprint is published by the registered company Springer Nature Singapore Pte Ltd.
The registered company address is: 152 Beach Road, #21-01/04 Gateway East, Singapore 189721, Singapore

Preface

This book entitles *Service Excellence for Sustainability—Lessons from Malaysia, Japan, and Taiwan* contains a collection of articles written by a keynote and invited speakers at the 2nd Service Science Symposium held on 24th November 2019. This book provides a significant conceptual and empirical contribution to understanding service science and service sectors' practices in the wake of industrial revolution 4.0. There are seven chapters in this book. The chapter contributors discuss the service economy's growth and how it remains sustainable in the current service trends. Multiple real-world cases of service systems from Malaysia, Japan and Taiwan are presented in this book.

The Service Science symposium is a symposium organised by the Serviceology Society Malaysia (SESMA), established on 29th July 2018. The society aims to bring service researchers and practitioners under one roof to redefine service concepts and ideas and explore the application of these concepts and ideas in real industrial and community settings. The word "Serviceology" used in naming the society was first coined by the Japanese Serviceology Society, where "-ology" is used for creating new logic. It explores the theoretical structures of service concepts and generates scientific systematisation of services and products from the transdisciplinary science and technology viewpoints, promoting technological developments in solving industrial and social issues.

Serviceology Society Malaysia (SESMA) vision is to lead society in service science and service innovation to strengthen Malaysia's service economy. The society mainly focuses on the co-creation of service value by facilitating effective networking and collaboration among the university, industry, and community members to achieve a sustainable Malaysian service economy. With this vision, we strongly hope that this book would promote knowledge sharing and emphasise service science essential and accelerate innovation in various research fields on services utilising a common theoretical understanding. And not just that, we hope this book will draw attention from scholars and practitioners not only from Malaysia, Japan and Taiwan but from the Asia-Pacific region too.

We want to thank the chapter authors for their contributions to this book. The names and short biographies of the authors are in the respective chapters. We also would like to thank Springer for providing us with the opportunity to publish this recent work. Special thanks to the reviewers for contributing their generous time and expertise in giving us constructive feedback. Finally, we wish to express special appreciation to our senior colleague, Assoc. Prof. Dr. Muriati Mukhtar, who encouraged and motivated us to prepare this book.

Bangi, Malaysia

Nur Fazidah Elias
Ruzzakiah Jenal
Hazilah Mohd Amin
Hazura Mohamed
Siti Aishah Hanawi

Contents

Contributors

Nur Fazidah Elias Faculty of Information Science and Technology, Universiti Kebangsaan Malaysia, Bandar Baru Bangi, Malaysia

Mohammad Rezal Hamzah Centre of Excellence for Social Innovation and Sustainability, Faculty of Applied and Human Sciences, Universiti Malaysia Perlis, Arau, Malaysia

Siti Aishah Hanawi Faculty of Information Science and Technology, Universiti Kebangsaan Malaysia, Bandar Baru Bangi, Malaysia

Anas Hasbullah School of Technology Management and Logistics, Universiti Utara Malaysia, Changlun, Malaysia

Huzili Hussin Centre of Excellence for Social Innovation and Sustainability, Faculty of Applied and Human Sciences, Universiti Malaysia Perlis, Arau, Malaysia

Adila Ismail Centre of Excellence for Social Innovation and Sustainability, Faculty of Applied and Human Sciences, Universiti Malaysia Perlis, Arau, Malaysia

Ruzzakiah Jenal Faculty of Information Science and Technology, Universiti Kebangsaan Malaysia, Bandar Baru Bangi, Malaysia

Farah Azea Khalid Faculty of Social Sciences and Humanities, Universiti Kebangsaan Malaysia, Bandar Baru Bangi, Malaysia

Fu-Ren Lin Institute of Service Science, National Tsing Hua University, Hsinchu, Taiwan

Hazura Mohamed Faculty of Information Science and Technology, Universiti Kebangsaan Malaysia, Bandar Baru Bangi, Malaysia

Hazilah Mohd Amin Faculty of Information Science and Technology, Universiti Kebangsaan Malaysia, Bandar Baru Bangi, Malaysia

Norani Nordin School of Technology Management and Logistics, Universiti Utara Malaysia, Changlun, Malaysia

Adam Mohd Saifudin Operation Management Department, School of Technology Management and Logistics, College of Business, Universiti Utara Malaysia, Changlun, Malaysia

Kunio Shirahada Japan Advanced Institute of Science and Technology (JAIST), Nomi, Japan

Yazrina Yahya Faculty of Economics and Management, Universiti Kebangsaan Malaysia, Bandar Baru Bangi, Malaysia

Nizamuddin Zainuddin Logistics Management and Transportation Department, School of Technology Management and Logistics, College of Business, Universiti Utara Malaysia, Changlun, Malaysia

Chapter 1
Service Science Sustainability from the View of Asia Pacific

Nur Fazidah Elias, Ruzzakiah Jenal, Hazilah Mohd Amin, Hazura Mohamed, and Siti Aishah Hanawi

Abstract The service definition is changing throughout the times. Services were previously defined as work done by a person to benefit others, such as banking, health, transportation, and hospitality services. However, today's world, laden with information, communication, and technology, has made traditional services into service science. Service science focuses on service systems that provide value for users to be satisfied with the service innovations received. Thus, the service system's efficiency and quality can be improved to create value in the service. The existence of value occurs when consumers consider the advantages and disadvantages of a product or service. Consumer involvement in the development of service produces the value co-creation. Through the value co-creation, the company as a service provider and the consumer as the service recipient interact to perform joint activities and create value based on the consumer experience. The interaction of consumers and companies to co-create value is service innovation. This chapter is an introduction to the discussion of service innovation in the real world. An overview of service science, value co-creation, and service innovation are discussed in detail. A summary of each chapter on issues in service excellence and sustainability is also discussed.

N. F. Elias · R. Jenal (✉) · H. Mohd Amin · H. Mohamed · S. A. Hanawi
Faculty of Information Science and Technology, Universiti Kebangsaan Malaysia, Bandar Baru Bangi, Malaysia
e-mail: ruzzakiahjenal@ukm.edu.my

N. F. Elias
e-mail: fazidah@ukm.edu.my

H. Mohd Amin
e-mail: hazilah@ukm.edu.my

H. Mohamed
e-mail: hazura.mohamed@ukm.edu.my

S. A. Hanawi
e-mail: ctaishah@ukm.edu.my

N. F. Elias et al. (eds.), *Service Excellence for Sustainability*,
https://doi.org/10.1007/978-981-16-2579-4_1

1.1 The Beginning

Services have long been part of the world economy. Traditionally, services are differentiated from goods by contrasting the tangible characteristics of goods with the intangible attributes of services. Classic examples of services include banking, healthcare, transportation, consulting, hospitality, and entertainment. However, with the advancements in information, communication, and operations technologies, the capacities and scope of services have been extended. They are continued to be extended beyond this traditional definition. In 2004, foreseeing a brave new future for services and its transformation into its organisation, International Business Machines (IBM) made a move to introduce a new transdisciplinary area named service science, management, engineering, and design, or service science for short. To date, service science is an established academic discipline and has developed into a highly researched area. Vargo and Lusch (2004, 2008) provide service science with a theoretical underpinning by the formulation of service-dominant logic.

The dominant service logic comprises ten foundational premises, each articulating the essence of service proposed by service science. Under this logic, service is always the fundamental basis of exchange. The dominant service logic also removes the dichotomy between goods and services. In this logic, goods are just vehicles or mediums for service provision. In foundational premises, the word service, in its singular form, is defined as applying resources and competencies to benefit others. Thus, service includes all economic activities. Individuals, organisations, and technologies work together, using specialised competencies and capabilities to create the most value. Under this new notion, an organisation is not proposed to be consumed by its customers; however, it is co-created by all parties involved to achieve something of value. Hence, service is value co-creation.

1.2 Service Science

Service science focuses on service systems and produces value to obtain consumer satisfaction and create a systematic basis for service innovation. Service science is the observation, involvement, knowledge of business, and the use of technology to analyse service systems, enhance innovation, and create value (Stoshikj et al., 2016). Service science is needed to transform and improve service efficiency and quality to create value in service.

Service science was introduced by IBM in 2005 and focuses on services consisting of science, management, and engineering aspects (Huo & Hong, 2013). The service system is a core element of in-service science that uses various resources, skills, and knowledge that involve competition so that a change can be made to create value. The value is created when consumers consider the advantages and disadvantages of purchasing products or using services (Siltaloppi & Vargo, 2014).

Service science uses systematic search principles to understand and improve value creation (Vargo & Akaka, 2012). Services are fundamental in the value co-creation. The value co-creation concept in-service science explains the methods and provision of services by using existing resources to create value.

Service science for the value co-creation consists of four types of resources: individual, technology, organisation, and information (Maglio & Spohrer, 2013). All four categories of resources are essential because they include resources with rights (individuals and organisations), resources as property (technology and information), physical entities (individuals and technology), and socially constructed entities (organisations and information). Such resources become the basis and entity in the formation of value co-creation.

1.3 Value Co-creation

The value co-creation is a theory that involves consumer collaborations in the development of services (Galvagno & Dalli, 2014). It is also a value that involves emotional relationships, positive findings, and knowledge sharing (Payne et al., 2009). Value co-creation is defined as the value created between the company and the consumer. It involves activities and interactions to integrate operations based on available resources to achieve value (Bettencourt et al., 2014). The interaction of value co-creation occurs when both parties are actively involved in value creation, influencing each other and interacting directly in the technological environment.

Previously, the concept of value co-creation was company-centred. Traditionally, the company has no interaction with the consumer because it is not seen as a value co-creation source (Prahalad & Ramaswamy, 2004). Thus, the company-centred concept faces consumer satisfaction and declining profit (Taghizadeh et al., 2016). The traditional concept is a one-way interaction: the company provides the services or sells the products. At the same time, the consumer receives services or products.

The concept of value co-creation is used to create an interaction between the company and the consumer to allow the activities to be done together and create new value. The revolution of interaction took place in line with the increase in digital technology. The company's interaction with consumer emphasis on consumer power because the value created is based on consumer engagement and experience through the provided platform (Ramaswamy, 2011). Therefore, the company needs to have an interactive platform to enable the interaction to take place continuously to create value.

The concept of value co-creation is essential because it prioritises the consumer. The value determined is based on the consumer's own experience after using the services. The success or effectiveness of service can also be measured through consumer satisfaction. The practice of value co-creation can increase consumer satisfaction when consumers can participate in the service and produce value (Hoyer et al., 2010). The consumer plays a vital role in improving a company's performance by telling others about their products, services, or brands, including new services to co-create experiences and values (Brodie et al., 2011). The consumer has different expectations and readiness levels while participating in value co-creation (Hakanen & Jaakkola, 2012).

The value co-creation model considers the aspects of value in use and is a process that requires interaction and collaboration between the company and the consumer besides the engagement and co-creativity. Thus, the value co-creation model should consider the processes that occur in consumers' use and their relationship to the consumer's objectives. The value co-creation models discuss here are DART Model, Payne's model, and Sampson's model called Process-Chain-Network.

1.4 Value Co-creation Models

Prahalad and Ramaswamy (2004) introduce a value co-creation model known as the DART model. This model involves interaction relationships between companies and consumers. The DART model has four main factors: dialogue, access, risk assessment, and transparency, as shown in Fig. 1.1. Dialogue (D) is a process that involves interaction, interactivity, and a tendency to act. This factor focuses on the method of communication between the company and the consumer. Access (A) starts with information and equipment that allows users to access data and information. This factor enables users to access the same information without being hindered by digital gaps. Risk (R) assessment involves the user's actions and decisions after being exposed to the possible dangers of using it. This factor makes consumers aware of the potential risks they need to face if they want to use a service. Transparency (T) is an explicit action and strategy implemented by the company to gain consumers' trust and confidence. Nowadays, the company no longer hides information such as prices, costs, and profits to reassure consumers that all service information is transparent and detailed. Factors in the DART model can also be combined and adapted in different ways to enhance user interaction. According to Maduka (2016), the best concept for value co-creation can be produced if all DART model factors are met.

The Payne model is a co-creation framework that considers product-related issues and shared experiences about products or services (Payne et al., 2009). The Payne model has four main components: the customer's value-creating process—concerned

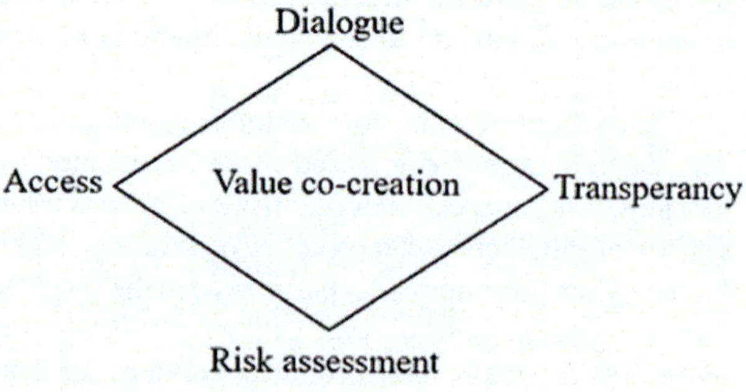

Fig. 1.1 DART model (*Source* Prahalad and Ramaswamy [2004])

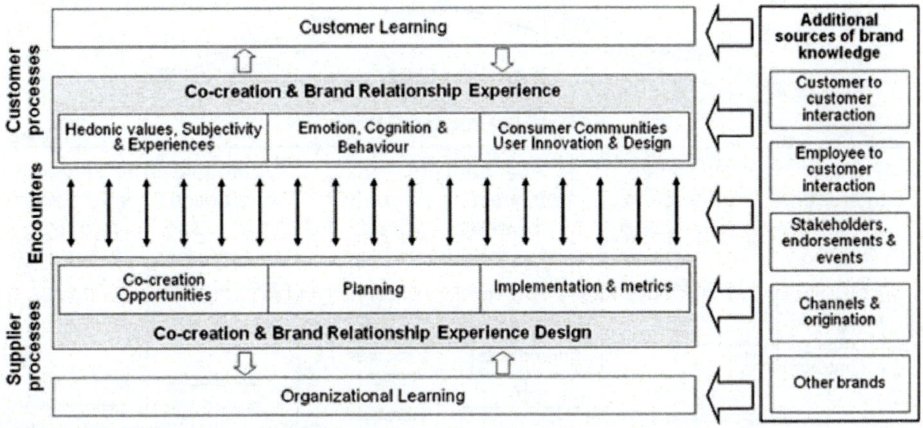

Fig. 1.2 Payne's model (*Source* Payne et al. [2009])

with co-creating and experiencing a brand relationship; the supplier's value-creating process—concerned with designing and co-creating a brand relationship experience; encounters where ongoing interactions are involved in creating these experiences; and, the impact of additional sources of brand knowledge (Keller, 2003) (Fig. 1.2). Payne model aims to identify the concept of relational experience for product or service in value co-creation and service-dominant logic, which produces a conceptual model for the design and management of user experience. Payne et al. (2009) describe how the value co-creation model is used to design, plan, and manage relationship experiences. Thus, through Payne model, the information obtained can explain how service development management and customers are involved with co-creation. Through the Payne model, the relationship between service providers and service recipients can be seen more clearly. The role between the two parties in how they create value can also be seen.

The Sampson model or Process-Chain-Network (PCN) is a value co-creation model that balances assessing communication aspects between users and service providers (Sampson, 2012). Through PCN, the complexities of a process that involves engagement between service providers and service recipients can be systematically analysed. According to Grönroos (2011), PCN is a 'production' in adding value in the traditional form of providing resources to achieve needs. Therefore, the available resources have a potential value. An example of a PCN diagram at the pizza restaurant is shown in Fig. 1.3.

PCN is simply a sequence of processes with an identifiable purpose to increase an entity's well-being or set of entities, creating value co-creation (Grönroos, 2011). PCN emphasises the processes that occur to manage service operations where the entity requirements, entity resources, and procedures implemented by each entity are effectively stated. PCN can identify complex service operations as a result of the processes involved between networks and entities. PCN also shows the domains involved between service providers and service recipients that lead their respective focus. The Sampson model provides a concrete definition of service with direct

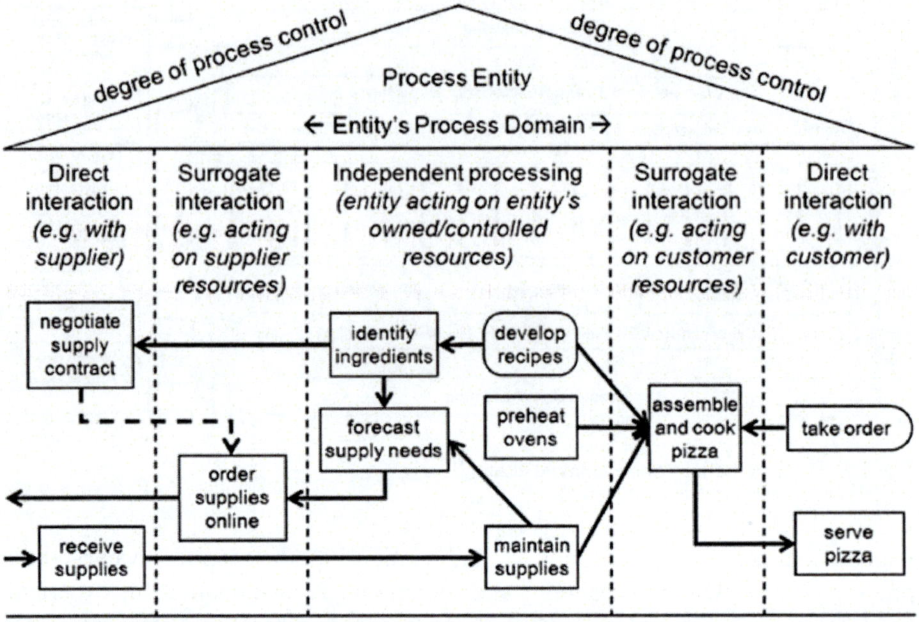

Fig. 1.3 Process chain network diagram. Example—at the pizza restaurant (*Source* Sampson [2012])

interaction steps and overlapping interactions in the domain process. Therefore, the processes that occur need to be studied in detail because they involve the entire process chain for a product or service, such as traditional operations, consumer needs, and service provider chains (Sampson, 2012).

1.5 Service Innovation

Services is an intangible artefact as opposed to tangible goods. Services is an area of intense research and has been for a long time. Indeed, the growth of service has been tied with the development of information and knowledge. However, in 2004, the emergence of Service Science was initiated by a telephone conversation between Jim Spohrer, who was starting up the IBM ® Research Service Research department, and Henry Chesbrough, a business and innovation professor at the University of California at Berkeley. The motivations and the deliberations by IBM in proposing this initiative are documented in several research papers (Chesbrough & Spohrer, 2006; Spohrer & Kwan, 2009; Spohrer & Maglio, 2009).

Chase and Apte (2007) gave a fascinating evolutionary account of service operations, tracing the big ideas in service operations from the 1900s until the 2000s, culminating in acknowledging the "Service Science, Management and Engineering" emergence initiative by IBM. The emergence of service science as a new interdisciplinary area of study addressed the challenge of becoming more systematic about innovating services. Service science involved the research, design, and evolution of service systems, which is defined as a value-creating configuration of people, technology, value propositions, and shared information. As a discipline, service science will uncover methods and approaches to study, design, evaluate existing service systems, and subsequently discover innovation points. It is these methods or techniques that will help industries to innovate and evolve systematically.

It thus seeks to answer questions such as how can service systems grow and achieve economies of scale? How is the fragility and robustness of service systems, including people and technology, similar and different from naturally occurring physical, chemical, and biological complex systems? To answer these questions and more, understanding the various service components present in our society is needed. In this book, the book's contributors discuss the service economy's growth and how organisations remain sustainable in current service trends. Multiple real-world cases of service systems are presented in this book. We briefly summarise each chapter in Sect. 1.6.

1.6 Issues on Service Excellence and Sustainability

Chapter 2, by Fu-Ren Lin, presents a real-world case of service system ecosystem development for achieving sustainable development of urban and rural service systems in Hsinchu, Taiwan. The service-dominant logic and service innovation practice are adopted to sustain the urban-rural value co-creation relationship. The project covers Jianshi Township, Hengshan Township, and the border between Zhubei City and Qionglin Township. In Jianshi, the sustainability issues faced are mainly on maintaining the Atayal culture with sustainable economic development. In Hengshan, the Jiuzantou Station on Neiwan Rail Line is revitalised by reopening the passenger hall. While in the border between Zhubei and Qionglin, the potential value of co-creation between city dwellers and countryside farmers is explored. This project brings Zhubei dwellers to experience the agricultural value by strengthening their relations with farmers and purchasing local agricultural products. Thus, the city dwellers and countryside farmers can co-create value that sustains their living quality and well-being. The cooperative service systems for the governance structure and management of the aggregation of resources from urban and rural areas are proposed. It is a crucial issue faced by expanding the city and preserving countryside agricultural value to achieve urban-rural sustainability.

Chapter 3, by Kunio Shirahada, consists of five topics. The first topic is about service research as a new institutional co-creation paradigm. Service research has been evolving and focusing on building relationships with customers. In the early 2000s, the service-dominant logic was introduced, which greatly affected the service research paradigm. Service had been considered a value creation process, and the importance of value co-creation between participants was recognised. The service ecosystem is a crucial part of service research. In the second topic, the author focuses on the transformative service economy and society. Nowadays, it is vital to take care of the well-being of the people. This has influenced transformative service research since society's well-being is closely related to service research. As the service economy grows, it gives enormous side effects to the natural ecosystems. Therefore, the service economy must be more sustainable to protect future generations from being compromised. The third topic covers the conservation and activation of service capital. Service is cardinal in economic exchange. Operant resources have been identified as the salient factor in ensuring sustainability. Kunio Shirahada recommends service to be treated as meta-capital as service capital will promote other capitals' sustainability. The fourth topic is about interests enabled by digitalisation and networking. An object is seen by the number of interests in it, which will determine its status. The value of service heavily depends on the situation at that time. Digitisation and networking are fundamental in promoting interests. The last topic is toward the new service research paradigm. There are three important points for the future institutional co-creation paradigm: the relationship between institutions and relativism, the rationality standard of co-creation, and actors' involvement in the natural ecosystem. In conclusion, the current service research paradigm is the sustainability paradigm. It is essential to consider the new base for the rationality of co-creation based on existing resources while keeping in mind the existence of others and future generations.

Chapter 4, by Adam Mohd Saifudin and Nizamuddin Zainuddin, focuses on the logistic issues in Malaysia. Economic liberalisation and globalisation have prompted Malaysia to participate in international trade aggressively. Inland ports in Malaysia play a significant role in the freight transport system providing access for producers and manufacturers in the hinterland to seaports acting as gateways to the nation's trade. The current inland ports in services in Malaysia include Padang Besar Cargo Terminal (PBCT), Ipoh Cargo Terminal (ICT), Segamat Inland Port (SIP), and Nilai Inland Port (NIP). These inland ports' existence facilitates containers' movements to major seaports (i.e., Port Klang, Penang Port, and Port of Tanjung Pelepas) and goods across the borders (Tebedu). This chapter reviews the logistical importance of an inland port in Malaysia. As for that matter, this chapter had collected various experiences of developing inland ports, reviewed their facilities and functions, and thus highlighting their importance. The findings suggest that inland port will play a more strategic and creative role in managing the demands and making the inland port an essential hub with the One Belt One Road (OBOR) initiative.

Chapter 5, by Farah Azea Khalid, and Yazrina Yahya, discusses sustainability and how to design and develop mobility programme at higher education institutions. As part of the internationalisation of higher education missions, many universities are putting effort into establishing platforms for their students to go overseas and provide similar programmes at home university for an exchange programme with their university collaboration partners. The mobility programme is an opportunity for students to gain more education, extracurricular, and socialisation experience. Quoting from the authors, "a sustainable programme will certainly promote the local students' socio-economic development and assist in providing better quality education," this chapter focuses on designing and developing a sustainable mobility programme that meets the global community. The chapter addresses two key questions; (1) how these mobility programmes can be sustainable, and (2) why these mobility programmes need to be sustainable. The authors expand the "Model for Sustainable Mobility Programme" by enhancing sustainable solutions. The new model provides a better elaboration of the solutions with detailed processes for executing the solutions. Farah and Yazrina implemented the model at UKM-GSMP, a mobility programme conducted by UKM that was initiated ten years ago. They end the chapter by describing how the expanded model would help a mobility programme be sustained for an extended period.

Chapter 6, by Huzili Hussin, Adila Ismail, and Mohammad Rezal Hamzah, discusses empowerment. By empowering peoples' sensibility that their voice matters and making a difference, social media technologies have changed how people communicate and act online. Understanding media psychology values and principles are crucial. Media psychology uses psychological concepts and theories to analyse and understand the complicated relationship between human experience and media technologies. Thus, it is very wise to know how media and communication technologies may impact our lives' psychological aspects, maintain the positive impacts, and minimise (or eliminate) the adverse effects it will bring. So that will increase the sustainability of communication services.

Chapter 7, by Anas Hasbullah and Norani Nordin, discusses servitisation. Manufacturers and business owners are forced to re-examine their business models to survive the onslaught of globalisation and reduce profits. They looked to services and servitisation as a source of their competitive advantage. The concept of servitisation was first coined in the late 1980s by Vandermerwe and Rada (1988). It then became a manufacturing industry strategy that discovered that their traditional practice of selling pure products is no longer sustainable (Belal et al., 2012). The concept can be defined as a combination of products and services where manufacturers offer "bundles" of services and products to add value to their existing products. This is also known as an "integrated product and service offering" that manufacturers offer to their customers. The study on servitisation is prevalent and growing fast in developed countries. However, in developing countries like Malaysia, it is a new concept. Therefore, it is essential in understanding the drivers that will enable Malaysia's industries to adopt servitisation. This chapter underlines the necessary groundwork in this direction by exploring the critical success factors for adopting the servitisation strategy. It is discovered that there are six essential success factors of adopting the servitisation strategy, which are vision, organisation structure, employee management, strategy, value chain, and customer empowerment.

1.7 Conclusion

Service is the application of resources and competencies to benefit others. Service science evaluates existing service systems through observation, involvement, knowledge of business, and the use of technology and subsequently discover points for innovation to transform and improve service efficiency and quality. In the subsequent chapters, real-world cases of service system ecosystem development show how various organisations incorporate service science for achieving sustainable development.

Acknowledgements This work is supported by Universiti Kebangsaan Malaysia through the Research Grant GGP-2020-038.

References

Belal, H. M., Shirahada, K., & Kosaka, M. (2012). Knowledge space concept and its application for servitizing manufacturing industry. *Journal of Service Science and Management, 5*(June), 187–195.

Bettencourt, L. A., Lusch, R. F., & Vargo, S. L. (2014). A service lens on value creation: Marketing's role in achieving strategic advantage. *California Management Review, 57*(1), 44–66.

Brodie, R. J., Hollebeek, L. D., Jurić, B., & Ilić, A. (2011). Customer engagement: Conceptual domain, fundamental propositions, and implications for research. *Journal of Service Research, 14*(3), 252–271.

Chase, R. B., & Apte, U. M. (2007). A history of research in service operations: What's the big idea? *Journal of Operations Management, 25*(2), 375–386.

Chesbrough, H., & Spohrer, J. (2006). A research manifesto for services science. *Communications of the ACM, 49*(7), 35–40.

Galvagno, M., & Dalli, D. (2014). Theory of value co-creation: A systematic literature review. *Managing Service Quality: An International Journal, 24*(6), 643–683.

Grönroos, C. (2011). Value co-creation in service logic: A critical analysis. *Marketing Theory, 11*(3), 279–301.

Hakanen, T., & Jaakkola, E. (2012). Co-creating customer-focused solutions within business networks: A service perspective. *Journal of Service Management, 23*(4), 593–611.

Hoyer, W. D., Chandy, R., Dorotic, M., Krafft, M., & Singh, S. S. (2010). Consumer co-creation in new product development. *Journal of Service Research, 13*(3), 283–296.

Huo, J., & Hong, Z. (2013). Chapter 2: The rise of service science. In V. Hugo (Ed.), *Service science in China* (pp. 39–68). Springer.

Keller, K. L. (2003). Brand synthesis: The multidimensionality of brand knowledge. *Journal of Consumer Research, 29*(4), 595–600.

Maduka, O. B. (2016). Effects of customer value co-creation on customer loyalty in the Nigerian service industry. *International Journal of Business and Management, 11*(12), 77–82.

Maglio, P. P., & Spohrer, J. (2013). A service science perspective on business model innovation. *Industrial Marketing Management, 42*(5), 665–670.

Payne, A., Storbacka, K., Frow, P., & Knox, S. (2009). Co-creating brands: Diagnosing and designing the relationship experience. *Journal of Business Research, 62*(3), 379–389.

Prahalad, C. K., & Ramaswamy, V. (2004). Co-creating unique value with customers. *Strategy & Leadership, 32*(3), 4–9.

Ramaswamy, V. (2011). It's about human experiences and beyond to co-creation. *Industrial Marketing Management, 40*(2), 195–196.

Sampson, S. E. (2012). Visualising service operations. *Journal of Service Research, 15*(2), 182–198.

Siltaloppi, J., & Vargo, S. L. (2014). Reconciling resource integration and value propositions: The dynamics of value co-creation. In *Proceedings of the Annual Hawaii International Conference on System Sciences* (pp. 1278–1284).

Spohrer, J., & Kwan, S. K. (2009). Service science, management, engineering, and design (SSMED): An emerging discipline-outline & references. *International Journal of Information Systems in the Service Sector (IJISSS), 1*(3), 1–31.

Spohrer, J., & Maglio, P. P. (2009). The emergence of service science: Toward systematic service innovations to accelerate co-creation of value. *Production and Operation Management, 17*(3), 238–246.

Stoshikj, M., Kryvinska, N., & Strauss, C. (2016). Service systems and service innovation: Two pillars of service science. *Procedia Computer Science, 83,* 212–220.

Taghizadeh, S. K., Jayaraman, K., Ismail, I., & Rahman, S. A. (2016). Scale development and validation for DART model of value co-creation process on innovation strategy. *Journal of Business & Industrial Marketing, 31*(1), 24–35.

Vandermerwe, S., & Rada, J. (1988). Servitization of business: Adding value by adding services. *European Management Journal, 6*(4), 314–324.

Vargo, S. L., & Akaka, M. A. (2012). Value co-creation and service systems (re)formation: A service ecosystems view. *Service Science, 4*(3), 207–217.

Vargo, S. L., & Lusch, R. F. (2004). Evolving to a new dominant logic for marketing. *Journal of Marketing, 68*(1), 1–17.

Vargo, S. L., & Lusch, R. F. (2008). Service-dominant logic: Continuing the evolution. *Journal of the Academy of Marketing Science, 36,* 1–10.

Dr. Nur Fazidah Elias is a Senior Lecturer at Faculty of Information Science and Technology, Universiti Kebangsaan Malaysia (UKM), Malaysia and a senior researcher at e-Service lab, Centre for Software Technology and Management, UKM. She received her B.CompSc. and M.Eng. from Universiti Teknologi Malaysia (UTM) and obtained her Ph.D. in Information Systems from Queensland University of Technology (QUT), focusing on IS success or IS impact. She specialises in information system success, and her research interests include the impact of IS/ES on organisations, IS cultural studies, system user satisfaction, service quality, e-service quality, survey design and validation. She has published several articles in top-rated conferences and journals in those areas.

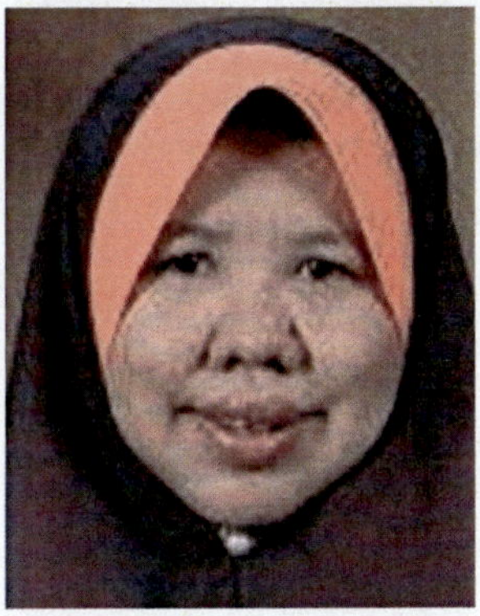

Dr. Ruzzakiah Jenal received her Bachelor of Information Technology in Industrial Computing from Universiti Kebangsaan Malaysia (UKM), M.Sc. in Industrial Mathematical Modelling from Loughborough University and Ph.D. from UKM. Her Ph.D. is in Mathematics, focusing on the area of optimisation of scheduling. She is currently a Senior Lecturer at Faculty of Information Science and Technology, UKM, Malaysia, and a researcher at e-Service lab, Centre for Software Technology and Management, UKM. Her research interests are in mathematical modelling, statistical modelling, service sciences, service delivery and e-service quality. She has published several articles in conferences and journals in those areas.

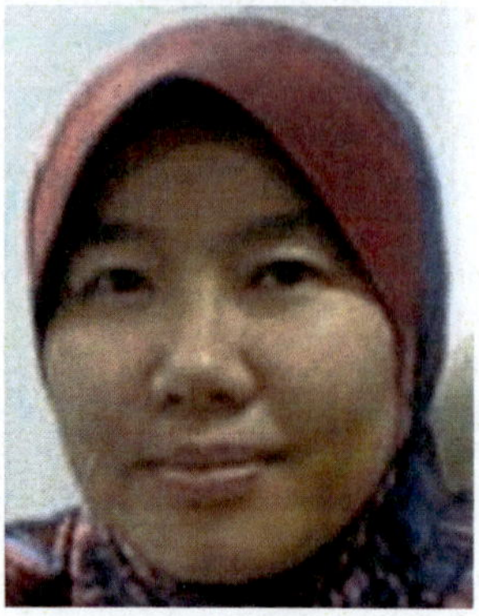

Hazilah Mohd Amin is a Senior Lecturer at Faculty of Information Science and Technology, Universiti Kebangsaan Malaysia (UKM), Malaysia. At UKM, she has taught numerous courses such as Entrepreneurship, Statistics, Simulation, Computer Modeling, Management Science, Information Technology, Management Information Systems, and Accounting Information Systems. She has published several articles in the *Asia-Pacific Journal of Information Technology and Multimedia, Jurnal Teknologi (Sains Sosial), International Education Studies, Journal of Social Sciences*, and other related research papers. She holds a Bachelor Degree in Mathematics and Statistics from the University of Adelaide, Australia and Master of Business Administration from Ohio University.

Dr. Hazura Mohamed received her Bachelor of Mathematics from Universiti Kebangsaan Malaysia (UKM), M.Sc. in Quality and Productivity Improvement from Universiti Kebangsaan Malaysia (UKM) and Ph.D. in Mathematics from Universiti Teknologi Malaysia. She is currently a Senior Lecturer at Faculty of Information Science and Technology, UKM, Malaysia, and a researcher at e-Service lab, Centre for Software Technology and Management, UKM. Her research interests include service quality, data analytics, satisfaction model and quality model. She has published several articles in conferences and journals in those areas.

Siti Aishah Hanawi is a Senior Lecturer at Faculty of Information Science and Technology, Universiti Kebangsaan Malaysia (UKM), Malaysia. She received her Bachelor of Science (Mathematics) and Master of Information Technology (Industrial Computing) from Universiti Kebangsaan Malaysia (UKM). She is also a researcher at e-Service lab, Centre for Software Technology and Management, UKM. Her current research interests are in mathematical and statistical modelling, service sciences, exploratory data analytics and quality models. She has published several articles in conferences and journals in the related areas.

Chapter 2
Service Science and Sustainability

Fu-Ren Lin

Abstract Service science is one of the interdisciplinary studies of service systems established in the early twenty-first century. It aims to tackle the value creation issues in complex systems with Service-Dominant Logic in human societies heavily affected by the interactive effects of technological innovation and social changes. The academic research paradigm shifts toward an engaged scholarship to encourage the iterative cycle of discovering issues faced in the real world and generating solutions to solve the problems, facilitating theory building and technology development. In this chapter, the service science discipline is elaborated and connects its research activities to achieve sustainability goals. A case study is used for explaining the issues, involved stakeholders, methods, and tools to attain sustainability. The major contribution of this case study is to position the emerging discipline of service science onto the important missions to achieve the sustainability of human societies by modelling the complexity of service ecosystems as actor-to-actor networks using Service-Dominant Logic and innovating service systems that sustain the value co-creation processes among different stakeholders (actors) including the natural environment. This case study results serve as an example that could inspire other practices in similar service ecosystems to realize their sustainability.

2.1 Common Language for Service Science

Service science, an umbrella term to define an interdisciplinary study of service systems, including domain knowledge in service management, design, and engineering, was first proposed by IBM in 2004. This emerging discipline investigates the phenomena identified in transforming human societies from agricultural, industrial to service industries and reshaping the academic boundaries in tackling issues involved in service ecosystems, such as service innovation enabled by technological advancement (Spohrer & Maglio, 2009). In the 2012 IBM annual report, IBM profit was generated 45% from software, 41% from service, and 14% from hardware and

F.-R. Lin (✉)
Institute of Service Science, National Tsing Hua University, Hsinchu, Taiwan
e-mail: frlin@iss.nthu.edu.tw

N. F. Elias et al. (eds.), *Service Excellence for Sustainability*,
https://doi.org/10.1007/978-981-16-2579-4_2

financing, comparing the proportion of 27, 38, and 35%, respectively, in 2000. It indicates the trend of IBM's profit shifts toward service and software, which is also the trend of many companies.

The world economic development moves toward a service economy, as shown in Table 2.1. The Global Domestic Product (GDP) contributed by service is 65% in 2018 as agriculture, industry, and manufacturing took 3, 26, and 16%, respectively. It implies the trend of demanding interdisciplinary human capital and domain knowledge in the service economy era.

2.1.1 Service System View

Service science then was defined commonly as the study of the complexity of service systems, in which a service system is defined as a value co-creation configuration of people, technology, internal and external systems connected by value propositions, and shared information (such as language, laws, measures, model, etc.). Sometimes, service systems and service ecosystems are used interchangeably. The Amazon Kindle service's service ecosystem is an example to elaborate on service systems' complexity to elaborate on service science concerns. Amazon launched the Kindle e-book service to deliver a digital copy to its customers who may use the Kindle reader as the reading device to read e-books sold in the Kindle store in 2007. The Kindle reader costs $399.

In 2009, Amazon started to sell Kindle (2nd generation) for international customers to download e-books from the Kindle store worldwide at the same price as in the USA. The Kindle reader cost $299 in 2009. The price of the Kindle reader is $69. If the Kindle reader is treated as a product, a manufacturer faces the challenge that the Kindle reader's price drops dramatically, which means that the device's market value is depreciated. However, if the formulation of the Kindle service system is treated, the service ecosystem is sustainable as the Kindle store is the largest online store globally, which serves readers worldwide.

Figure 2.1 shows the composition of stakeholders in the Kindle service system. Amazon owns Kindle store, which encodes the proprietary digital format of e-books from publishers, and then leverages the different interfaces such as desktop, notebooks, iPad, and smartphone to grant the accessibility for readers to read contents purchased from the Kindle store via internet connections with Wi-Fi or mobile data telecommunication by telecom carriers of the world. It also leverages social media, such as Facebook, to ease readers to share their reading items within their social networks, promoting Kindle store. The Kindle store is a service system composed of different stakeholders, including people, technology, and internal and external systems that contribute various resources for the sustainability of value co-creation. The value proposition issued by the Kindle store is to grant users' flexibility to access e-books on different reading devices with the same price around the world. A reader has his/her agency to select the contents and interfaces to enjoy the value delivered,

Table 2.1 The world development indicator

	Gross domestic product		Agriculture		Industry		Manufacturing		Services, value added	
	$ billion		% of GDP		% of GDP		% of GDP		% of GDP	
	2010	2018	2010	2018	2010	2018	2010	2018	2010	2018
World	66,036.90	85,804.40	4	3	27	26	16	16	63.2	65
East Asia & Pacific	16,935.10	25,918.30	6	5	36	34	24	23	57	59.8
Europe & Central Asia	20,947.20	23,041.50	2	2	24	23	14	14	64.1	64.3
Latin America & Caribbean	5,347.20	5,787.30	5	5	29	24	14	13	55.8	60.3
Middle East & North Africa	2,768.10	3,616.90	5	5	46	39	14	14	47.4	54
North America	16,611.30	22,213.30	1	1	20	18	12	11	75.2	77.4
South Asia	2,061.10	3,457.80	18	15	29	26	16	15	46.6	49.9
Sub-Saharan Africa	1,369.70	1,699.50	16	15	28	24	9	10	50.7	52.4
Low income	368.4	5733	27	26	24	25	8	8	40.2	39.3
Lower middle income	4,462.00	6,707.50	16	15	32	29	16	15	46.5	49
Upper middle income	15,856.60	24,435.10	7	6	37	33	22	21	50	55.3
High income	45,352.00	54,118.80	1	1	24	23	14	14	69.1	69.8

Source http://wdi.worldbank.org/table/4.2#

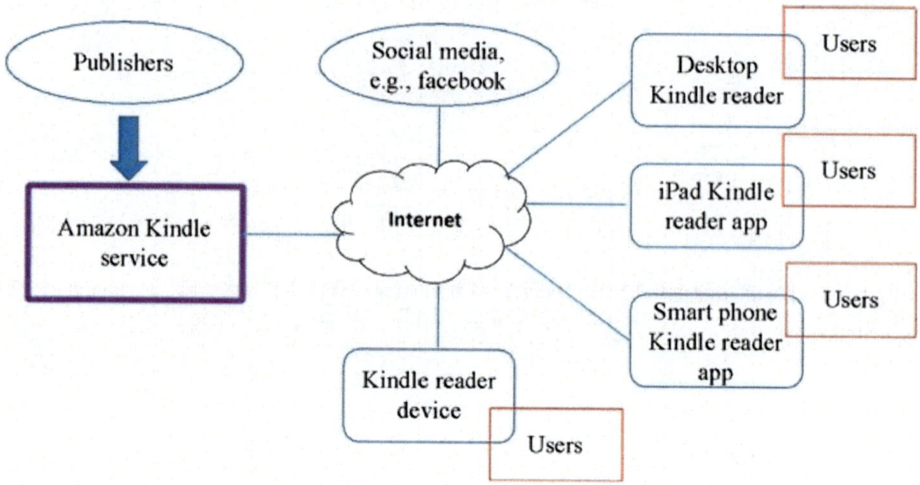

Fig. 2.1 Amazon Kindle service system

through which the service provider and the user co-create the value of the established service ecosystem.

From the example of the Kindle service system, the Kindle reader serves as a medium for readers to access Kindle e-books, which are the value perceived by readers. In contrast, readers have the cognitive ability to digest the contents. It indicates that both sides of value exchange need to spend their resources to create value from this service. The Kindle service system's formation exemplifies the Service-Dominant Logic proposed by Vargo and Lusch (2008) and summarized in Lusch and Vargo (2014).

2.1.2 *Service-Dominant Logic*

The evolutional composition of Service-Dominant Logic (S-DL) is organized, as shown in Fig. 2.2. It composes four axioms expanding to 10 foundational premises. In S-DL, service is the fundamental basis of exchange (FP1), service exchanges for service. Service is the value one exercises one's operant resources, e.g., knowledge and skill, to benefit the other for receiving the retrospective value the other contributes (FP4). Thus, the value is co-created between a service provider and its customers (FP6). Since the value is co-created, an enterprise can only propose a value proposition to attract its customers to co-create value with it (FP7). Thus, the service-centred view of a service system is customer-oriented and relational (FP8). The service system participants are economic and social actors who are resource aggregators to contribute to their viability (FP9). The value is then determined uniquely and phenomenological by the beneficiary (FP10), demonstrating S-DL's evidentially defining service and evaluating its value. In service exchange, goods

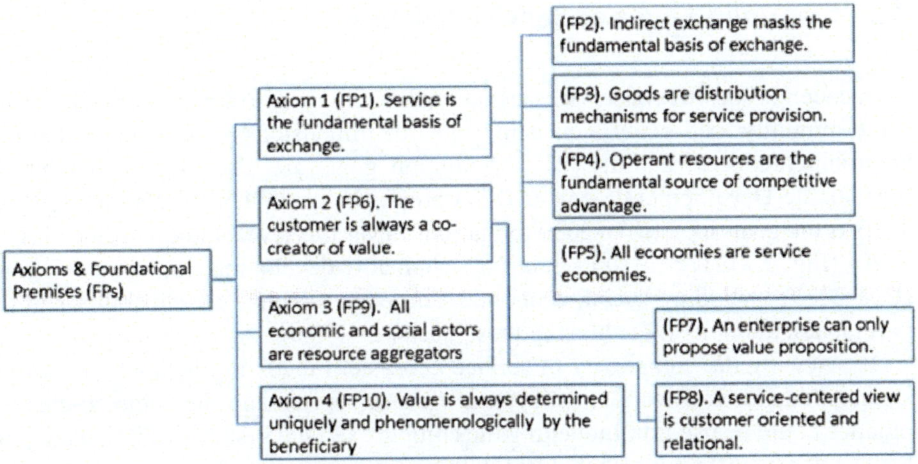

Fig. 2.2 Axioms and foundational premises of Service-Dominant Logic (*Source* Lusch and Vargo [2014])

serve as the distribution mechanism for service provision (FP3). That is what the Kindle device is acting.

The Amazon Kindle service example explicates the complexity of the e-book retailing service system with the S-DL. The Amazon website displays e-book information collection. A potential reader as the buyer can access the information via the Internet. The Amazon website can only issue a value proposition of corresponding book information that means something for viewers who become the buyer to purchase and download the e-book to his/her preferred devices, such as smartphone, iPad, or read it on notebooks via cloud service. At this stage, the value is co-created between the Amazon Kindle service and the buyer. The buyer is an actor who searches on the web to identify information and then aggregates Amazon's resources via Kindle service. The authors' knowledge is delivered as information via Kindle service as the service to exchange for the reader's absorption of the knowledge by exercising his/her cognitive abilities. That is a service exchange for service. From the authors' source of value to the value recipient, i.e., the reader, there is indirect exchange masked by the service participants in the service system; for example, the telecom carriers that enable Kindle users to download the content seamlessly. To view all participants involved in a service system as actors to co-create value for corresponding actors with S-DL, the service science community could understand and communicate phenomena identified in service systems and collaborate for service innovation across disciplines.

2.2 Common Goals of Service Science

The academic societies of service science commonly agree to develop service science for community well-being, consisting of three dimensions: economic, social, and environmental sustainability. These common goals are also emphasized by the Sustainable Development Goals (SDGs) proposed by the United Nations (UN) in 2015 as the goals to reach in 2030 for nations in various levels of organizations (UN, 2020) (Fig. 2.3). The 2030 agenda for sustainable development specifies 17 goals (UN, 2020), including social, economic, and environmental sustainability and the participatory actions for achieving these goals.

By viewing the complexity of service ecosystem involving interaction between artificial and natural worlds, the system view of service and the value co-creation practice could be used for modelling the complex service systems while taking joint actions across different stakeholders with various aspects and expertise in tackling challenges on the way to achieve sustainable development goals. This study will present an ongoing real-world case of service system ecosystem development to elaborate on the methodology and tools of service science for achieving sustainable development of urban and rural service systems in Taiwan.

Fig. 2.3 17 SDGs proposed by the UN (*Source* https://www.un.org/development/desa/dspd/2030agenda-sdgs.html)

2.2.1 A Case Study of Urban–Rural Sustainability

In this chapter, an engaged scholarship model is connected with the real-world case in urban–rural sustainable development in Hsinchu, Taiwan. First, the engaged scholarship model is elaborated; second, a cyclic model of the product-service system (PSS) is used for service innovation to revitalize the local economy while sustaining social and environmental well-being. Third, continuous efforts are introduced to fully adopt the S-DL and service innovation practice to sustain the urban–rural value co-creation relationship on revitalizing a region in Hsinchu County. Finally, the lessons learned are summarized from these endeavours to articulate service science with sustainability.

2.2.2 Engaged Scholarship for Service Science

Service science is centred on humanity, which demands the domain knowledge related to service management, technology, and design. It emphasizes the design-led innovation to fulfil the well-being of societies. For academics, the science of service relies heavily on the iteration of problem formation, theory building, research design, and problem-solving activities that is coherent with the outside-in and inside-out process of service design and innovation. The process starts to discover the real-world problems, define and understand the problems, innovate problem solutions, and then put the solution into practice to sustain the service system. Figure 2.4 elaborates the iterative process of practising engaged scholarship (Van de Ven, 2007).

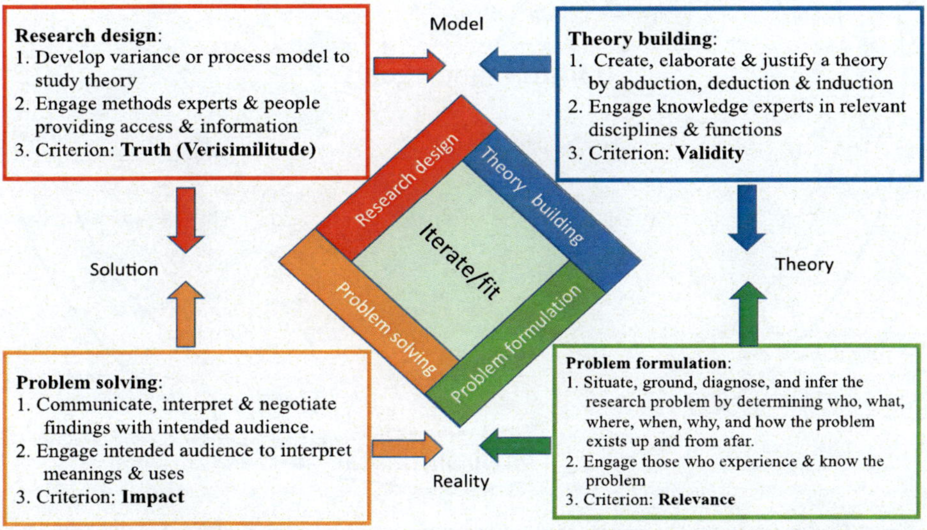

Fig. 2.4 The iterative process of practicing the engaged scholarship (*Source* Van de Ven [2007])

2.2.3 A Cyclic Model of Product Service System (PSS)

Xing et al. (2013) developed a cyclic model of the product-service system to deal with synergistic community transformation by integrating systems theory and product-service system (Fig. 2.5). In Fig. 2.5, the circle presents four dimensions of concern in transforming communities, which are social and political, cultural, economic, and environmental consistent with what SDGs demand. The cycle of transformation starts from sensing the target field's real-world situation, e.g., urban and rural communities. The stakeholders involved in the service ecosystem treated as partners (viewed as actors in service systems) are formed as collaboration networks. Different actors with competencies across disciplines are organized as capability networks to aggregate resources (e.g., social capital, financial capital, natural capital, etc.) to co-create value to change the status of the quo of the service system following the scenarios that envision the outcomes of the innovated service systems from delivering the implemented product-service solution to the field. The feedback from the field will

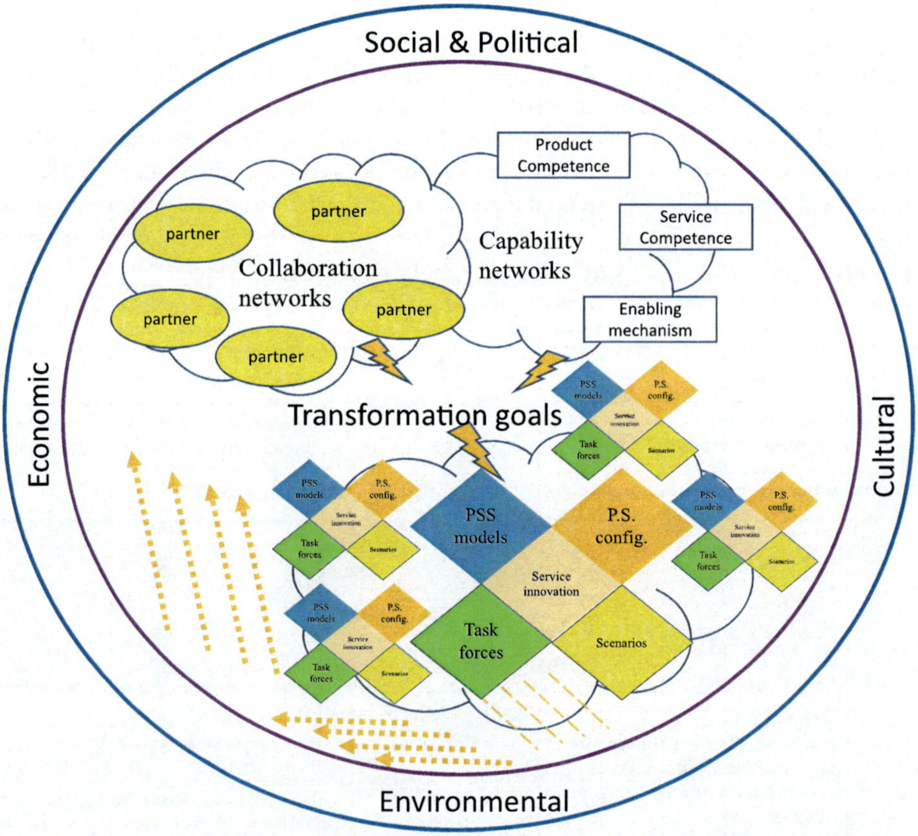

Fig. 2.5 A product-service system view of service innovation (*Source* Xing et al. [2013])

continue the cyclic process of service system innovation. The case study of urban–rural sustainability via service innovation has been adopting the PSS model's cyclic process.

2.2.4 A Case Study of Urban–Rural Sustainability via Service System Innovation

2.2.4.1 The Background Information of the Case Region

The PSS model's cyclic process shown in Fig. 2.5 was derived from the target region's field study, Hengshan Township Hsinchu County, 20 kilo meters east from Hsinchu City (Xing et al., 2013). Figure 2.6 shows Hsinchu City and the country's geographical regions, including the townships that Route 120 passes through from Zhubei City, Qiongli Township, Hengshan Township to Jianshi Township. It composes Satochi, Satoyama, and Okuyama areas in the east of Hsinchu City, extending from the urban living zone, agricultural townships, and mountain townships.

The Youluo River flows from Jiangshi Mountains, passes through Hengshan, Qionglin townships, and then merges with Shanping River as Touqian River flowing into the Taiwan Strait. With the forests in Jianshi Mountains, the mountains had contributed to the lumbering industry ended 30 years ago; the coal and cement mining industries also took advantage of the Satoyama area in Hengshan Township ended

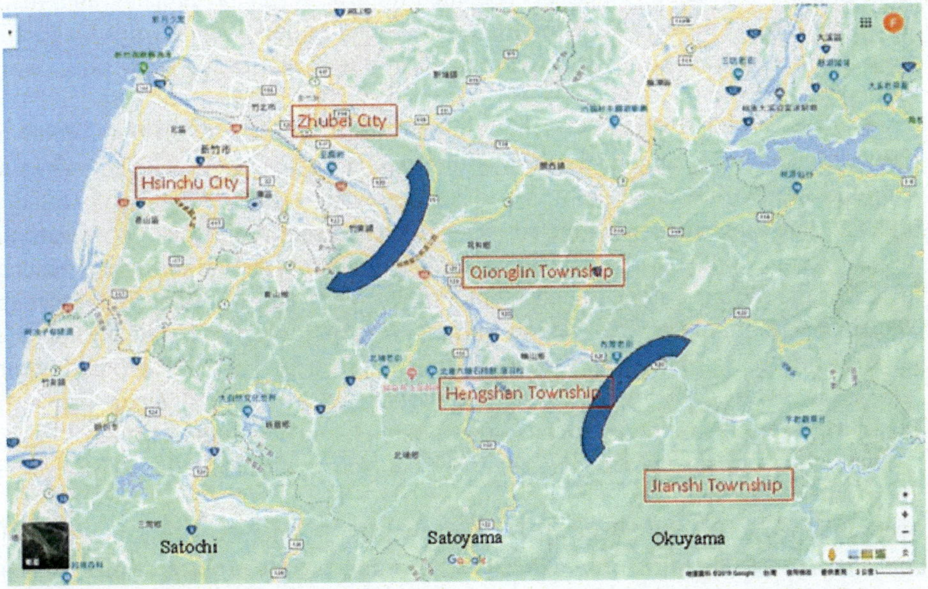

Fig. 2.6 The geographic distribution of urban–rural regions in the case study (*Source* Snapshot from Google Map and marked by the author)

twenty years ago. The local communities of these townships collapsed along with the end of the lumbering and mining industries. A majority of workforces were flowing to work for factories in Hsinchu industrial zones. This region's reduced population is mainly due to industrial transformation that reduced working opportunities within this region and the increasing job opportunities created from the science park in Hsinchu City. The push-out and suck-in effects from rural and urban areas respectively creates issues, such as the appreciation of housing price in cities, the demand of recreation for urban residences, the competition in educational resources in urban while the lack of sufficient intellectual resources in rural areas, and the idle commercial business in the rural area suffering community economy while the city dwellers are suffering bad air quality and competing limited shared public services, which lower the living quality and well-being for people both in rural and urban areas.

By viewing the region status as the outcomes of the interactions among different existing service systems, in past years, we followed the S-DL. We adopted service design methods and tools to change the status quo by motivating involved actors to mobilize public and private resources to initiate service innovation projects.

2.2.4.2 Regional Revitalization Process

The involvement of service science with the regional revitalization process started in 2008 when the course, Introduction to Service Science, took Hengshan Township as the target area for students to understand the interactions of different service systems that rendered the economic status, social, and environmental conditions. The process of regional revitalization actions is specified in three stages: awareness, alignment, and integration. These three stages are not the end of regional revitalization's continuous actions; however, it helps explain the continuous journey so far has been gone through.

Awareness: In the awareness stage starting from 2008, the new master program on Service Science was launched at the Institute of Service Science in the National Tsing Hua University, Taiwan. The new view of service systems and the value of co-creation methods and tools were investigated, which led the research and teaching toward the new directions: outside-in and inside-out. Thus, the Introduction to Service Science course is the stepping stone to understanding service systems' complexity. Since the social and economic service systems in rural areas of Taiwan were not as structured or servicetised as urban areas, they could bring students opportunities to understand the issues involved with the existing stakeholders in creating values to sustain the service ecosystems involved social, environmental, and economic aspects.

In this stage, residences in Hengshan Township are connected to identify concerns they paid most attention via contextual inquiries. Since 2008, students in the class were grouped to several teams visited the township to interact with residences and specified topics that they were interested in forming as term projects to propose new service systems to solve regional sustainability problems. Students were equipped with a design thinking mind-set and tools for service innovation. There have been such issues as K-12 educational resources, agricultural sustainability, natural environment protection, community healthcare service for aged residences, and the economic growth in the region. Among proposed innovated service prototypes, some prototypes have been recognized as potentially implementable new services.

1. Bike service systems operated by communities. The Neiwan rail line has been run for years, and there are five stations in Hengshan Township. However, most local stores closed due to the lack of connections between the tourists and the nearby residential communities due to customer base shrinkage. Even worse, the tourists passing through Route 3 have been satisfied by the convenience stores, such as 7–11 and Family Mart, and they seem not to need any other store services. Thus, it is essential to identify the unmet demands of the visiting tourists and innovate new services. A new bike service operated by the communities was proposed that enables tourists from train stations and bus stops to explore the communities with natural and historical resources, which could create new business opportunities for local commerce.
2. A cooperative platform to bring up agricultural value from local small farmers. The agriculture of Hengshan Township benefited from the clean water and fertile earth in the past hundred years. However, due to the nearby cities' industrialization, the second generation of farmer families left for other living opportunities thanks to the better education, which caused many dispelled farms in this region. For those who continue farming, mainly aged farmers, it is challenging to create commercial value due to the lack of channels to trade their produces. A cooperative platform serving as a service system was proposed to connect the value created by local farmers to create a joint channel to shop produces to the local residences' trading posts; moreover, these farmers could offer urban dwellers experiential services, the natural-agricultural value.
3. An open museum as a service ecosystem. By Hakka culture with sustainable natural and agricultural resources, the ongoing service innovation projects could become the periodical and continuous events that could occur according to seasons and cultural calendar. The curation of events following service design and innovation practices co-creates the value of regional sustainability with involved public and private stakeholders, which define the value of the open museum: The museum of Dashanbei, taking the hill of Dashanbei as the landmark of the open museum (Fig. 2.7).

Alignment: In the alignment stage of regional sustainability, starting in 2012, there are two breakthrough accomplishments besides the continuous course projects. The first accomplishment is the inclusion of four professors from different disciplines and universities in Hsinchu to form joint projects sponsored by the Service Society of Taiwan to innovate regional sustainability services. One project is to enhance the STEM education for local public education by engaging external educational foundations and students' parents. The other project is to engage a public bus service system to connect the hill of Danshanbei with tourists mainly via Neiwan Rail Line. The second accomplishment is revitalizing a dispelled primary school campus on the hill of Dashanbei to become a centre for visitors to experience local history, agriculture,

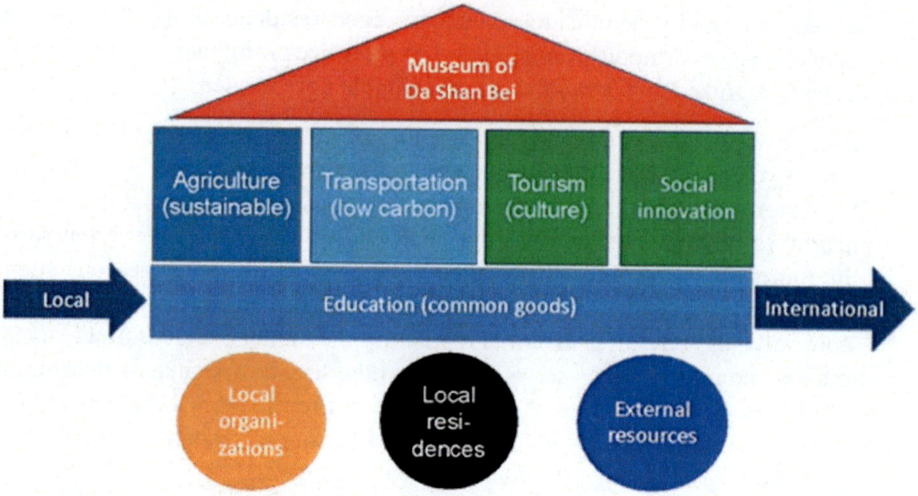

Fig. 2.7 The service ecosystem for the Museum of Dashanbei

and business service. The revitalized campus was called "Dashanbeiyue," representing the four emphases of the Museum of Dashanbei, Sky, Earth, Humanity, and Ecology. An entrepreneur who graduated from the Institute of Service Science started up the innovative service system to co-create value with local agricultural systems and educational systems to generate new value propositions to achieve sustainability in natural, agricultural, and cultural aspects market-driven business transactions.

Integration: The integration stage of regional sustainability started in 2017 as the Ministry of Education (MoE) and the Ministry of Science and Technology (MoST) started to pay attention to the social impact of universities on the regional revitalization by sponsoring projects for universities to integrate their multiple disciplines to enhance the sustainability of local communities and industries. The MoE in 2017 started to extend the involved stakeholders in Hengshan Township to formally propose the Museum of Dashanbei service ecosystem to the public to motivate the general public to join the effort on sustaining the regional well-being. The project granted by the MoST under the direction of human innovation and social practice started in 2018 to integrate different disciplines in National Tsing Hua University to conduct social innovation-related research projects to sustain the societies along Route 120, which highlight the urban–rural sustainable development issues (Fig. 2.5).

The project covers three geographical areas containing specific issues, respectively. The area in Jianshi Township mainly dwelling Atayal people who withhold traditional Gaga tribal leadership (a type of community norm to maintain Atayal's tribal societies) orally carried through generations by tribal leaders. However, due to the increasing interactions with Han people, who are the major population in Taiwan and also hold the major economic resources in the society, the Atayal people face the challenges in preserving their culture and living intelligence that their ancestors maintained the sustainability in social, economic, and environmental aspects. The sustainability issues faced in this area are mainly on maintaining the Atayal culture with sustainable economic development. Outside communities in Taiwan can co-create value with Atayal people to enjoy the diversity and inclusion of multiple cultures.

The Hengshan Township project aims to revitalize the Jiuzantou Station on Neiwan Rail Line by reopening the passenger hall. Space is renovated to serve as the visitor center for the Museum of Dashanbei to connect local residences and the tourists who could take the train or drive to explore nature and experience innovative services. The close of the train station passenger hall also kept the distance between two communities nearby the station. Thus, the project works on two integrated tasks. One task is to connect the local communities and visiting tourists via the curation of local history exhibitions. Local residences serve as the storytellers and artefact collectors to facilitate the exhibition curated by a professor in history. The other task is to enhance the art education in the local schools and communities, so that the organizations' living contexts will be gradually enriched by the art design; moreover, the inspiration of art creation will extend the value co-creation among the local residences, residential artists, and visitors. These tasks' touch-points' evolutional outcomes will be accessed via the information centre located at Jiuzantou Station passenger hall.

The border between Zhubei City and Qionglin Township explores the potential value of co-creation between city dwellers and countryside farmers. Due to the increasing population of Zhubei City that newcomers are mainly workers in the Hsinchu Science Park companies, the city border has been expanded in past years to absorb new movers. It caused the shrinking land size once was rice fields and farming villages. This project deals with the upcoming wave of land planning hosted by the Hsinchu County Government to convert 300 hectares between Zhubei City and Qionglin Township to the commercial and residential zone. This is not a single case that happened in Taiwan due to urbanization and industrialization. Therefore, the issues faced by the city's expansion and the preservation of countryside agricultural value are important for sustainability. This project aims to bring Zhubei dwellers to experience the richness of agricultural value by strengthening their relations with farmers and purchasing local agricultural produce as their routine life to co-create city dwellers. Countryside farmers' value sustains their living quality and well-being.

2.3 Service Science for Sustainability

In the practice of service science for sustainability, the rural areas comparing with the service systems in the urban area is considered by the S-DL; for example, the population density in cities could sustain many types of commercial services, such as dining, recreation, cultural events, etc. which may provide job opportunities and economic activities. However, the over-commercialized urban service systems may not sustain the well-being and good quality of life. Meanwhile, urban service systems' absorption power causes the decreasing rural population, severely destroying existing rural service systems and decreasing rural areas' resource density to sustain service systems.

Take the Hsinchu case in this chapter as an example. The regional revitalization process's integration stage aims to aggregate the resources inherited and owned by involved actors from urban and rural areas to design and innovate regional service systems to co-create value for people in both areas. That is what the Museum of Dashanbei aims to achieve to preserve the natural resources, innovate service offerings by integrating resources from various actors to co-create value, and sustain the service ecosystem via a viable governance structure. That is also the regional revitalization project sponsored by the MoST to increase each township's resource density on Route 120 and then innovate service systems that can fulfil the demands from urban dwellers and achieve regional economic, social and environmental sustainability.

Based on the service system view and value co-creation of service science, the cooperative service systems for the governance structure and management of the aggregation of resources from urban and rural areas are proposed to achieve regional sustainability. A cooperative is defined by the International Cooperative Alliance Statement on the Co-operative identified as "… an autonomous association of persons united voluntarily to meet their common economic, social, and cultural needs and aspirations through a jointly-owned and democratically-controlled enterprise."

The suitability of developing cooperative service systems in the Hsinchu region could be justified as follows: (1) culturally fit: the agricultural communities have been established by cooperation in exchanging tools, labour, and produces to fulfil their needs; thus, it fits well with the Hakka and Atayal communities; (2) motivationally drive: the lack of various resources needed in innovated service systems drives the motivation for cooperation; and (3) accessible market: the urban dwellers in Hsinchu and Zhubei cities look for the good quality of life when they mainly work for tech companies. The variety of cultural, natural, and agricultural richness of the rural areas is easily accessible once rural areas' innovated service systems are established effectively.

The block chain technology will facilitate the cooperative service systems developed in the MoST project for realizing the cooperative principle: jointly owned and democratically controlled enterprise. Since block chain technology ensures the transparency of transactions and the decentralized operation, the cost of cooperative management could be reduced to increase the benefit of cooperative members, such that the value generated from the cooperative service systems could match the expectations from the urban dwellers who have been living in a relative capitalism business context.

The block chain-enabled cooperative service ecosystem is under development. Once it is formed and operated, the actors involved could effectively co-create value via innovative service processes and information exchange among actors via smart contract mechanism embedded in block chain technology. Moreover, the cooperative service ecosystem is enabled by the decentralized peer-to-peer information infrastructure, which hosts business processes. It ensures each actor's autonomy to enrol the value co-creation networks; the ecosystem could strengthen the resource aggregation among small farmers, store owners, artists, entrepreneurs, and residences to enhance the resource density that can drive the business processes to engage urban dwellers with the rural service systems.

To solidify the cooperative service ecosystem, the development of community currency, called "Coop Coin" will be facilitated to serve as the exchange currency among cooperative actors in the ecosystem to boost up the resource exchange among them which may accumulate the social capitals and also draw the flow of capitals from urban areas; then, the increase of capitals in the ecosystem strengthens the capabilities from local communities for urban–rural sustainability. The Museum of Dashanbei is included in the cooperative service ecosystem as a value proposition for sustainability.

The experiences learned from these three stages of regional revitalization (awareness, assignment, and integration) demonstrate the essence of democracy that enables the autonomy of local communities to seek their sustainability and the system view and value co-creation perspectives of service science that facilitate the methods and technologies to achieve sustainability. The sustainability of urban–rural sustainability could be a fertile field to practice engaged scholarship since the complexity of service systems demand very much interdisciplinary communication and collaboration, which is the core value of service science and innovation.

2.4 Conclusion

To tackle urban–rural sustainability issues, a case study as an example is presented to describe what service science is and how it could be used for modelling the service ecosystem involved in the urban–rural relationships and innovating new service systems for sustaining the viability of the urban–rural service ecosystem. This chapter's major contribution is to identify service science's core principles based on Service-Dominant Logic: system view of service and value co-creation of service innovation. Then, elaborate on how they have been used for service system innovation with the case study. The ongoing cooperative service ecosystem development based on block chain technology will significantly progress in technology-enabled service innovation for urban–rural sustainability. This chapter serves as a progressive report to remark the theories and practices of the status quo to inspire engaged scholarship.

Acknowledgements The author would like to express the gratefulness for receiving major sponsorship from the Ministry of Science and Technology, Ministry of Education, Service Science Society of Taiwan. The progress of urban–rural sustainability has been a joint effort from public sectors (townships, city and county governments), local community leaders, academic scholars, and many other participants from industries and communities. The author is grateful for their contributions to making the progress.

References

Lusch, R. F., & Vargo, S. L. (2014). *Service-dominant logic: Premises, propositions, possibilities*. Cambridge University Press.

Spohrer, J., & Maglio, P. P. (2009). The emergence of service science: Toward systematic service innovations to accelerate co-creation of value. *Production and Operation Management, 17*(3), 238–246.

UN. (2020, January 24). *The 2030 agenda of sustainable development*. https://sustainabledevelopment.un.org/?menu=1300.

Van de Ven, A. H. (2007). *Engaged Scholarship: A Guide for Organizational and Social Research*. Oxford University Press.

Vargo, S. L., & Lusch, R. F. (2008). Service-dominant logic: Continuing the evolution. *Journal of the Academy of Marketing Science, 36*, 1–10.

Xing, K., Ness, D., & Lin, F.-R. (2013). A service innovation model for synergistic community transformation: Integrated application of systems theory and product-service systems. *Journal of Cleaner Production, 43*, 93–102.

Dr. Fu-Ren Lin is a Professor and the founding director of the Institute of Service Science, National Tsing Hua University, Taiwan. He graduated from the University of Illinois at Urbana-Champaign in 1996 with a Ph.D. degree in Information Systems. He taught in the Department of Information Management, National Sun Yat-sen University until 2004. He joined the Institute of Technology Management in National Tsing Hua University until the Institute of Service Science was launched in 2008. He has been doing research in various topics, such as supply chain management, business process re-engineering, e-commerce, knowledge management, and data/text mining, since graduation. Based on the prior academic background and methodologies, he started to focus on service innovation-related issues in service science research, including new service development, healthcare service engagement, and service experience design, service value network analysis. His research results have been published in domain related journals, such as *Service Science, International Journal of e-Commerce, Decision Support System, Information Systems and e-Business Management, Information Processing and Management, Electronic Commerce Research and Application*, etc. He served as the general secretary for the Service Science Society of Taiwan in 2011–2015. He currently lead a research group sponsored by the Ministry of Education to work on the topic in business transformation and sustainability for resilience.

Chapter 3
Service Sustainability Paradigm as a Basis for Transformative Service Society

Kunio Shirahada

Abstract This chapter reviews the progress of service research to date and points out that the current research paradigm of service studies is based on "the new institutional co-creation paradigm". This current research paradigm should be modified to understand the future service economy and society which will be developed in the context of wellbeing, sustainability, and digitalization. Based on this background, this chapter proposes the concept of service capital as a meta capital to keep and improve accessibility to sustainability-related capitals such as environmental, economic, social one and so on. In addition, three additional perspectives necessary for the next-generation service research paradigm are also discussed; (i) the explanatory framework which maintains the diversity in the institution and the effect of digitization, (ii) understanding the other rationality standards of co-creation, and (iii) research framework for the involvement of actors in a natural ecosystem.

3.1 Service Research as a New Institutional Co-creation Paradigm

Service research has undergone a variety of changes since the 1980s (Fisk et al., 1993). Previous discussions have led to significant findings, such as the unique characteristics of services by comparison of services and goods. These findings were then applied to marketing strategies. Services were regarded as intangible actions for the purpose of providing benefits, with a focus on the quality of these actions. The actions became recognized as objects of trade, causing quality actions to become competitive. Service research was then focused on building relationships with customers to enhance the quality of the actions. Results showed the significance of the actions and the enrichment of customer experiences in processes of services.

At the beginning of the 2000s, Vargo and Lusch (2004) proposed service-dominant (S-D) logic as a world view, which greatly affected the service research paradigm.

K. Shirahada (✉)
Japan Advanced Institute of Science and Technology (JAIST), Nomi, Japan
e-mail: kunios@jaist.ac.jp

N. F. Elias et al. (eds.), *Service Excellence for Sustainability*,
https://doi.org/10.1007/978-981-16-2579-4_3

The concept of service came to be regarded as a value creation process via resource integration, and the importance of value co-creation between participants was recognized. Manufacturing industries have made efforts to understand the essence of service and apply it to their businesses to improve customer satisfaction and business model, a trend known as servitization. The motivation for resource integration is based on the maximization of individual rationality, also known as resource density (Frow et al., 2016), which establishes the so-called principle of economic rationality.

Another important concept in service research is the service ecosystem. A service ecosystem is defined as a relatively self-contained, self-adjusting system of resources and integrated actors that are connected by a shared institution and mutual value creation through an exchange of services (Lusch & Vargo, 2014). The various layers of the system affect service activities, particularly implicit and explicit institutions formed within the layers. The advantage of sharing institutions is stable activities, which has been studied in new institutional organization theory. In service research, this logic has been adapted as a service ecosystem, and studies have analysed the impact of institutions on co-creation activities such as the Actor-Institutions-Matrix (AIM) framework analysis (Baron et al., 2018).

Consequently, the paradigm presently established by service research can be regarded as the new institutional co-creation paradigm.

3.2 Perspectives on Transformative Service Economy and Society

What will become important in the service economy and society in the future? Around the world, various enterprises are investigating new working styles along with the use of artificial intelligence (AI) for the wellness of employees. In today's ageing society, governments are aiming to better society by creating opportunities for senior citizens to demonstrate their abilities and find their purpose in life while also ensuring their physical and mental health.

In a matured economy and society, it is essential to consider the well-being of people. The concept of well-being has been regarded as an important indicator of progress since transformative service research was first proposed in 2011 (Anderson et al., 2013) and has gradually continued to gain the attention of more scholars. As of November 2019, the number of management-related papers in the Web of Science that mentions transformative service research, service, and/or well-being has reached 217, with a significant increase over the past several years. Additionally, 4,199 citations were found for these topics (excluding self-citations).

Well-being is considered to include physical, psychological, and eudemonic aspects in relation to the fulfilment of one's potential abilities. It is generally defined as a state in which one finds meaning in life and maximizes one's abilities (Calvo & Peters, 2014). While well-being is a pluralistic concept for human beings, it is a value concept that human beings define intrinsically (Bradley, 2015). It is also compatible with the idea of transformative value (Blocker & Barrios, 2015), which claims that the value developed in S-D logic is habitual while the one developed in transformative service research is transformative. Unlike the values imposed on us from the norms and rules in a certain social structure, well-being is about positive and uplifting emotions based on intrinsic values, so it is not necessarily defined extrinsically in-group norms. The positive emotions include happiness from hope and optimism (Laros & Steenkamp, 2005). Hope is essentially a concept involving the passage of time. As the context surrounding oneself changes over time, it is important to form the belief that the world one desires is there even after time passes.

Regarding the temporal change in the context of the individual, one of the most important issues at present is climate change. Humans are wholly dependent on natural ecosystems (Shirahada & Fisk, 2014) which have been negatively affected by human activity since industrialization. In today's global service economy, service recipients include the more than seven billion people on the planet. As service economies grow, recipients are consuming more goods and services and putting greater pressure on the resources of the ecosphere, which will impact sustainability. If we look at the breakdown of electricity consumption that emits greenhouse gases, we can see that activities in the service sector as well as those related to human migration account for a large portion. As the service economy develops, electricity consumption and emissions will only continue to increase. On November 21, 2019, it was reported that the British band Coldplay decided to temporarily shelve their live tour activities in an effort to reduce the environmental impact of touring (Martin, 2019). The service sector needs to be more responsible for sustainability in order to combat environmental issues.

Sustainability, according to the EPA, is defined as "policies and strategies to meet the present needs of society without compromising the ability of future generations to meet their own needs" (EPA, 2010). The definition was later modified to include "creating and maintaining the conditions under which humans and nature can exist in productive harmony to support present and future generations" (EPA, 2019). Natural capital, human capital, social capital, and economic capital are all aspects that are related to sustainability (Goodland, 2002). Sustainability is closely linked to capital retention, which creates policies and determines assessments.

3.3 Conservation and Activation of Service Capital

According to Lusch and Vargo (2014), service is the fundamental basis of economic exchange by which value is created. Thus, services can be considered capital which is an important resource for sustainability. The key to making resources available is operant resources, i.e. skills and knowledge. Operant resources can be considered synonymous with human and cultural capital. In the case of service capital, it is possible to preserve the wisdom of human beings. Archiving, protecting, and making useful knowledge available are important for managing service capital in the era of sustainability.

In addition, this paper proposes to treat service as a meta-capital (Bourdieu & Wacquant, 1992; Driessens, 2013) which involves having power over various fields and forms of capital that circulate within them in order to access sustainability-related capitals such as environmental, economic, social one and so on. To make use of certain capital, the capital of its superordinate concept is required. For example, the concept of satoyama in Japan refers to a traditional system of rural land use that represents a balanced relationship between people and nature. The system is able to sustain a variety of ecosystem services and preserve the diversity of secondary natural environments (Takeuchi, 2010). Humans have acquired various ecosystem services while assigning value to natural ecosystems. The capital from services functions as the capital that creates the value of sustaining human and environmental capital. As another example, mobile catering activities for the elderly living in rural areas contribute to the formation of human and social capital by promoting the independence of the elderly and forming communities through long-term activities (Ho Bach & Shirahada, 2020).

Service capital as a co-creation process seems to have many possibilities as a motive force to promote the sustainability of other capital.

3.4 Interests Promoted by Digitalization and Networking

According to the economist Yuichi Shionoya, value is associated with interest and establishes certain conditions (Shionoya, 1984). The value of an object comes from the amount of interest in it. The interest gives the object a desirable status, and the value is an evaluation of the object with respect to the status. While economical, it also serves as a reminder for us to consider the relation between value and interest. In services, it is considered that the value determined by recognition of the actual existence renews the value judgment in relation to the amount of interest. For example, you become interested in a drink when you are thirsty on a hot day. If you are able to get a drink, then it becomes a meaningful object to you, giving you the factual condition of quenching your thirst and cooling you down. If the drink satisfies your conditions, it is considered valuable to you. In other words, the value of the drink is evaluated by considering the situation at that time. Then the value of the drink is judged, and a decision is made whether or not to drink.

Interest can stem from various sources, from one's deficiencies to social relationships, such as the desire to attract interest from others. In recent years, the spread of digitization has made it possible to increase the visibility of existence, giving us opportunities to raise new concerns. In addition, networking can potentially lead to the innovation of norms and rules by connecting those with the same interests. In Mahayana Buddhism, *pratītyasamutpāda* refers to the principle of interdependent causation. Something can exist, but human interest will change, and networks are formed. Service scholars can learn from Buddhism in considering the relationship among existence, interests, and value generation.

3.5 Toward New Service Research Paradigm

As mentioned previously, well-being is an important concept for the future of the service economy, as it correlates directly to the sustainability problem. More effective analytical perspectives need to be provided in terms of the diversity inherent in well-being and the potential of service capital inherent in sustainability. The following three points are suggested additions to the new institutional co-creation paradigm.

The first point is the relationship between institution and relativism. New institutional schools have discussed reducing trade uncertainty by sharing a common belief system, or an institution. Although the institution makes intra-system transactions robust, the possession of a common belief system ultimately leads to the homogenization of intra-system actors. This can lead to relativism, which separates us from other entities with different belief systems. On the other hand, since the new service economy and society are pluralistic, actors will not adapt to homogenization and will instead pursue new activities seeking diversity. The movement is also the driving force behind overcoming relativism. Therefore, it is necessary to examine the explanatory framework which maintains the diversity in the institution and the effect of digitization.

The second is the rationality standard of co-creation. The present paradigm is based on the idea that actors exchange and integrate resources and create value by sharing their abilities. The purpose of the rationality criterion is to maximize its resource density. Human rationality is related to reality; therefore, rationality cannot be brought about without external constraints. Presently, the framework for what can be grasped as reality seems to be limited to economics. At the same time, a sense of co-existence based on values generated by improvisation or even just gathering each other is functioning as valuable for the people in tourism cases. It shows the need to consider how the existence of others is perceived in individual co-creation rationality for well-being.

The third point is the involvement of actors in a natural ecosystem. The service economy bears a great responsibility for environmental sustainability. Though the ecosystem service concept has been introduced in environmental and ecological economics for some time now, discussions typically centre on the quantitative evaluation of monetary value. In the future, natural ecosystems will have a great impact on human decision-making and value judgment. Digitization enables natural ecosystem actors who have not spoken out to share their information digitally, accelerating the establishment of existence as equal actors with people in the economy and society. It is necessary to examine how to incorporate natural ecosystem actors into analyses of management behaviour and discussions of reality.

A sustainability paradigm is a new service research paradigm that will arise with the accumulation of cases and considerations from the aforementioned viewpoints. It is crucial to consider the possibility of freedom from homogeneous pressure is inherent, and new grounds for the rationality of co-creation, based not only on one's resource density but also on the existence of others and future generations.

Acknowledgements This work was supported by JSPS KAKENHI Grant Number 17K03925 and 20H01529.

References

Anderson, L., Ostrom, A. L., Corus, C., Fisk, R. P., Gallan, A. S., Giraldo, M., et al. (2013). Transformative service research: An agenda for the future. *Journal of Business Research, 66*(8), 1203–1210.

Baron, S., Patterson, A., Maull, R., & Warnaby, G. (2018). Feed people first: A service ecosystem perspective on innovative food waste reduction. *Journal of Service Research, 21*(1), 135–150.

Blocker, C. P., & Barrios, A. (2015). The transformative value of a service experience. *Journal of Service Research, 18*(3), 265–283.

Bourdieu, P., & Wacquant, L. J. D. (1992). *An invitation to reflexive sociology*. University of Chicago Press.

Bradley, B. (2015). *Well-being (key concepts in philosophy)* (1st ed.). Polity.

Calvo, R. A., & Peters, D. (2014). *Positive computing: Technology for wellbeing and human potential*. MIT Press.

Driessens, O. (2013). Celebrity capital: Redefining celebrity using field theory. *Theory and Society, 42,* 543–560.

EPA. (2010). *Basic information: Sustainability*. U.S. Environmental Protection Agency.

EPA. (2019). *Learn about sustainability*. U.S. Environmental Protection Agency.

Fisk, R. P., Brown, S. W., & Bitner, M. J. (1993). Tracking the evolution of the services marketing literature. *Journal of Retailing, 69*(1), 61–103.

Frow, P., McColl-Kennedy, J. R., & Payne, A. (2016). Co-creation practices: Their role in shaping a health care ecosystem. *Industrial Marketing Management, 56,* 24–39.

Goodland, R. (2002). Sustainability: Human, social, economic, and environmental. In *Encyclopedia of global environmental change* (pp. 1–3). Wiley.

Ho Bach, Q., & Shirahada, K. (2020). Barriers to elderly consumers' use of support services: Community support in Japan's super-aged society. *Journal of Nonprofit & Public Sector Marketing, 32*(3), 242–263.

Laros, F. J. M., & Steenkamp, J. E. M. (2005). Emotions in consumer behavior: A hierarchical approach. *Journal of Business Research, 58*(10), 1437–1445.

Lusch, R. F., & Vargo, S. L. (2014). *Service-dominant logic: Premises, perspectives, possibilities.* Cambridge University Press.

Martin, C. (2019, November 21). *Coldplay to pause touring until concerts are 'environmentally beneficial'*. BBC News.

Shionoya, Y. (1984). *The structure of values* (in Japanese). Toyo Keizai.

Shirahada, K., & Fisk, R. P. (2014). Service sustainability: A tripartite value co-creation perspective. In *Progressive trends in knowledge and system-based science for service innovation* (pp. 89–99). IGI Global.

Takeuchi, K. (2010). Rebuilding the relationship between people and nature: The Satoyama initiative. *Ecological Research, 25*(5), 891–897.

Vargo, S. L., & Lusch, R. F. (2004). Evolving to a new dominant logic for marketing. *Journal of Marketing, 68*(1), 1–17.

Dr. Kunio Shirahada is an Associate Professor in Japan Advanced Institute of Science and Technology (JAIST) from 2012. He received his Ph.D. at the University of Tokyo in 2009. His research interests include Transformative Service Research (service sustainability, community revitalization for collective well-being creation, service for the elderly), neuro-marketing using Near-Infrared Spectroscopy (NIRS) technology, motivation management for service employees, and innovative organization design. He has published in the *Journal of Service Theory and Practice*, *Technological Forecasting and Social Change*, *Technology in Society*, *Journal of Business Research*, *Int. J. of Technology Management*, *Int. J. of Automotive Technology Management*, *Service Industries Journal*, *Int. J. of Innovation and technology management*, and other Japanese service related research papers. He presented his paper at Frontiers in Service Conference, QUIS, SERVSIG conference, Transformative consumer research conference, and PICMET (Portland International Conference on Management of Engineering and Technology). He is a member of editorial board on IEEE-EM, and other technology management journals.

Chapter 4
Inland Port in Malaysia: Logistical Revisit

Adam Mohd Saifudin and Nizamuddin Zainuddin

Abstract Malaysia inland ports play a significant role in the freight transport system providing access for manufacturers and producers in the hinterland to seaports, as gateways to the nation's trade. The current Malaysia inland ports in services include Padang Besar, Ipoh, Nilai, Segamat and Tebedu to facilitate containers' movements to major seaports (i.e., Port Klang, Penang Port and Port of Tanjung Pelepas) and goods across the borders (Tebedu). This chapter's objective is to review the logistical importance of inland ports in Malaysia. As for that matter, this chapter had collected various experiences of developing inland ports, reviewed their facilities and functions, and highlighting their importance. The findings suggest that an inland port in the northern region of peninsular Malaysia is logistically relevant for an intermodal transportation hub in the Northern Corridor Economic Region (NCER), and a vital hub in line with the One Belt One Road (OBOR) initiative.

4.1 Introduction

The dry port is also known as the inland port (feeder port) is a part of a specialised centre or location developed to serve the intermodal transportation network. The definition of dry ports has evolved, particularly in terms of their functions. The most comprehensive definition by Roso and Lumsden (2009) describes dry ports as 'an inland intermodal terminal directly connected to seaports with high capacity transport means, where customers can leave and pick up their standardised units as if directly to a seaport'. A dry port could be an inland port within a country with a gateway port,

A. M. Saifudin (✉)
Operation Management Department, School of Technology Management and Logistics, College of Business, Universiti Utara Malaysia, Changlun, Malaysia
e-mail: msadam@uum.edu.my

N. Zainuddin
Logistics Management and Transportation Department, School of Technology Management and Logistics, College of Business, Universiti Utara Malaysia, Changlun, Malaysia
e-mail: nizamuddin@uum.edu.my

N. F. Elias et al. (eds.), *Service Excellence for Sustainability*,
https://doi.org/10.1007/978-981-16-2579-4_4

or it could be located in an adjacent land-locked district in the hinterland of one or more seaports. There are various purposes of developing dry ports or inland ports, which can be understood by their types in terms of distance such as close, mid-range and distance (Monios, 2011) and location including seaport-based, city-based and border-based (Beresford et al., 2012). Each type of dry port plays various roles in the freight transport system, such as a freight distribution centre, an extended gateway for seaports, and a hub of intermodal transport.

Inland port survival depends on the cargo shipping and logistics freight forwarding demands. It includes the asset-based facilities of the port to ensure the successful operation of the site. Thus, it is critical to the port's effective and efficient operational management to control financial and administrative costs. If a private entity manages an inland port, funding sources need to be secured before development begins so that the site acquires operation status. Marketing is an essential element in any inland port management plan. Without successful marketing, businesses will not know what to expect at a site or what benefits can be derived from locating at the inland port. Marketing staff must successfully attract businesses that will anchor the site and subsequently attract supporting businesses (Rodrigue et al., 2010).

Location is the crucial logistical factor in determining inland ports' success and their development parallel to the container seaports (Hanaoka & Regmi, 2011). The location of these facilities at an inland port relieves support for storage and customs space at the seaport itself. According to Jeevan et al. (2015), inland ports' location should be short from the manufacturing zone to create more ports' demands. Whether the distances between inland ports and seaports or other trade origins or destinations are short or long, certain principles must be applied in determining the location of inland ports. They are inland ports within a short delivery distance by road of trade sources, factories, warehouses or retail stores. This would imply that they should be located within or close to industry, manufacturing or logistics precincts or zones.

Infrastructures are needed to provide trans-loading activities in inland ports. Customers are generally willing to utilise the services provided by inland ports if good transport infrastructure and high standards and sophisticated equipment can be sufficiently equipped to handle their valuable containers safely. Besides that, to attract a more significant freight share, service quality and reliability must be improved, especially freight services' punctuality. In line with its role, an inland port includes facilities for storage and consolidation of goods, maintenance for road or rail cargo carriers and customs clearance services (Jeevan et al., 2015). The inland port needs to have an immediate site of the inland port long-term storage and repair facilities for trucks or rail wagon or locomotive maintenance. A generalised functional structure of an inland port is illustrated in Fig. 4.1.

Lower costs are essential for the firm to remain competitive with the product offerings. Logistic costs are a significant component of overall trade costs. Logistics costs are costs associated with the entire logistics chain payable by the cargo owners or shippers for local delivery; terminal handling and storage; line-haul transport (transport between inland ports and seaports); other intermediate costs (related to customs clearance). The customers' cost factor includes freight rates and charges for using the services, one of the main determinants in logistics costs. A reduction in

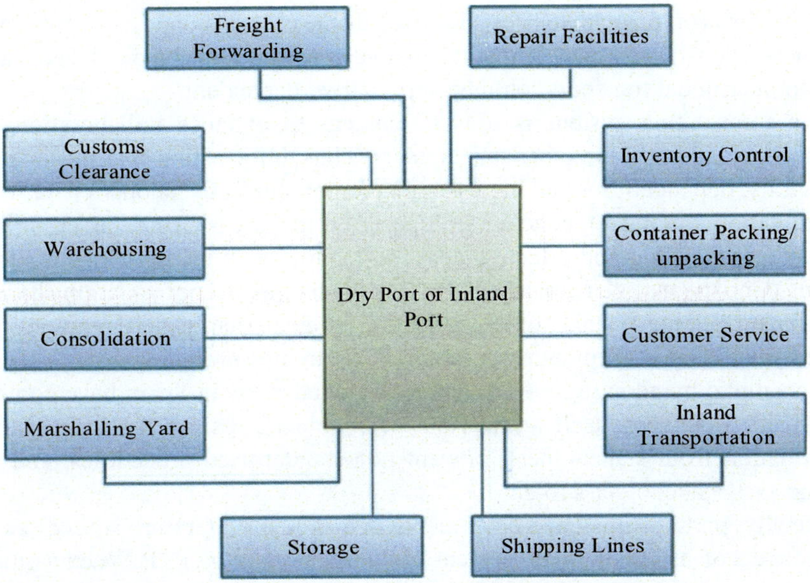

Fig. 4.1 Functional components of inland ports (*Source* UNCTAD [1991])

logistic costs may directly stimulate exports and imports. The decreased costs can make exports and imports more competitive (Cho, 2014).

Rapidly increasing foreign trade has generated a requirement for long-distance transport feeder services between major inland trade centres, which create seaports and trade demand areas—the mode of transportation consisting of the canal, pipeline, road, rail, air or vessel. Multimodal transportation is defined as the transportation of goods by using a sequence of two or more various modes of transportation (UNECE, 2009a, 2009b). Inland port should minimise the total transport cost. The railway can offer significant cost efficiency for freight haulage distance over 300 km. Therefore, inland port sites must be accessible by railways and highways networks with direct connections to seaports to minimise the total transport cost. Inland port must integrate railways with existing logistical networks to provide door-to-door service. Interfaces between railways and other transport modes are essential to encourage a modal shift. Changes in government policies regarding inland port transportation systems can help increase rail freight shares by developing rail freights as a business, encouraging a level playing field, private sector involvement and competition, and reducing barriers at borders.

The inland port concept is based on a seaport directly connected by rail with inland intermodal terminals where containers can be dealt with in the same way as in a seaport. Intermodal transport with inland ports could be a potential solution for seaport terminal congestion and better seaport inland access based on a short-haul rail. Intermodal transportation is generally considered viable on markets with more massive flows or longer distances. However, due to costs of congestion, growing environmental constraints, and competition between seaports, the break-even distance for inland intermodal transport could be very market dependent.

An information system is a vital element to enhance collaboration among seaport, inland ports and other stakeholders. Thus, information sharing can improve reliability, dependability and increase the momentum in the supply chain. Besides, coordination among stakeholders, especially in risk sharing, asset utilisation, accurate forecasting and effective decision-making is crucial for inland port operations (Christiaanse & Kumar, 2000). To enhance precise information sharing and coordination among stakeholders, and the efficient platform is required. In this regard, various port community systems (PCS) are mushrooming all over the world. For example, by adopting a PCS, Valencia inland ports in Spain have coordinated information between itself and various stakeholders, resolve the inconsistency of information from stakeholders, prevent non-standardised procedures, and clearly define each stakeholder's role.

Finally, partnerships have been mentioned in many literature sources as a vital component of successful management, including ports; e.g., PTP. When partnerships are created with metropolitan planning organisations, zoning and land-use changes can benefit operations at the inland port. Additionally, partnerships with local shippers' councils or professional organisations can help governmental agencies promote the inland port.

4.2 A Review: Inland Ports in Malaysia

The growing trade and container traffic in Malaysia has forced seaports to improve their capacities and services to support their business developments. Seaports in Malaysia seek inland port assistance to accelerate maritime trade momentum, increase intermodal application, improve seaport competitiveness, enhance regional economic developments, and establish Malaysian seaport policy (Jeevan et al., 2015).

In Malaysia, dry ports also known as inland ports (feeder ports), play a significant role in providing access for manufacturers and producers in the hinterland to seaports, which act as gateways to the nation's trade. In particular, inland ports in Malaysia have increasingly played an active role in facilitating the nation's trade, enabling goods to be transported and distributed from seaports to their final destinations. Malaysian inland ports have become essential logistics infrastructures for Malaysian container seaports to accommodate increasing throughput to remain competitive to attract customers from the foreland and hinterland. The transit points provided by inland ports enable the exporters and importers in the nation's hinterlands to connect with the nearest transhipment seaports. They also provide linkages between shippers/exporters and end-users of products/commodities.

Institutional issues are significant for the development of the transport service industry. Like most countries, Malaysia has a comprehensive legislative and regulatory framework governing both road and rail freight transport modes. The impacts on economic development and the sustainability of this framework are of great concern to the Malaysian government. Several laws govern the operation and services of road and rail transport. Due to the transport industry structure, several ministries and agencies are heavily involved with these policies and regulations. Coordination and integrations of all institutions related to logistics and transportation will ensure a robust, effective and efficient institutional framework established for the advantage of industry players and service providers (ports and inland port operators).

In Peninsular Malaysia, four inland ports are currently operating to support seaport container terminal operations. In line with this strategy, Malaysian container seaports are equipped with four inland ports, located perpendicularly connecting Thailand-Malaysia-Singapore. The first inland port in Malaysia is Padang Besar Cargo Terminal (PBCT), operating since 1984. Followed by three others at Ipoh, Nilai and Segamat have been developed since the 1990s to accelerate container movements to major seaports including Port Klang, Penang Port and Port of Tanjung Pelepas (PTP). Table 4.1 shows some available inland ports in Malaysia and their transportation networks.

Ipoh Cargo Terminal (ICT), the second inland port in Malaysia, was established in 1989. It is a mid-range inland port for Penang and a long-distance inland port for Port Klang and Pelabuhan Tanjung Pelepas (PTP). This inland port is connected with all Malaysia container seaports through road and rail links. ICT is a well-known inland port and helps import and export services for various industries in Northern Malaysia. The third Malaysian inland port is the Nilai Inland Port (NIP), a short-range inland port for Port Klang, and a long-distance inland port for PTP. NIP offers a range of

Table 4.1 Inland ports in Malaysia

Inland Ports	Location	Seaports connection
Padang Besar Cargo Terminal (PBCT)	158 km north of Penang Port and 588 km north of Port Klang	Road and Rail
Ipoh Cargo Terminal (ICT)	181 km south of Penang Port and 250 km of Port Klang	Road and Rail
Segamat Inland Port (SIP)	212 km south of Kuala Lumpur and 188 km north of Port Tanjung Pelepas	Road and Rail
Nilai Inland Port (NIP)	50 km South of Kuala Lumpur and 93 km from Port Klang	Road
Tebedu Inland Port (TIP)	Road Distance 1 km from the border of West Kalimantan, Republic of Indonesia and Sarawak, Malaysia 85 km from Kuching, Sarawak 100 km from Senari Container Terminal, Kuching, Sarawak 370 km from Pontianak, Republic of Indonesia	Road

Source Khalid (2012)

value-added services, facilities and spaces to support the growing container volumes at Port Klang in the central region and Johor Port in the south.

The Segamat Inland Port (SIP) commenced operation in 1998. Located in the southern part of peninsular Malaysia is a mid-range inland port for PTP and Port Klang. SIP offers facilities and services to manufacturers and traders in the southern region of peninsular Malaysia. The establishment of this inland port has made Johor Port and Port Klang the preferred ports of entry instead of neighbouring ports. SIP has been developed as a national load centre and transshipment hubs.

In Sarawak, there is one inland port called Tebedu Inland Port, which was established in 2004 to facilitate the cross border trade between Malaysia and Indonesia. Tebedu Inland Port caters for the international trade between Malaysia and Indonesia along the latter's Kalimantan border. The inland port in Tebedu in Sarawak facilitates not only the movements of goods across the borders but also people, hence contributing to promoting stronger bilateral ties between Malaysia and Indonesia.

Each inland port has significant roles and responsibilities to container terminals in Malaysia and the international transhipment of containers, providing feeder business to and from South Asia, Cambodia, Thailand and Vietnam. All Malaysian inland ports support container seaports' Trans loading activities by splitting the seaports' function in inland locations (Arvis et al., 2014). Stakeholders significantly demand functions such as interface nodes linking production zones with seaports, customs services, spaces for container storage and customised services. These inland ports also contributed to supply continuity of container volumes to seaports, providing additional capacity and acting as a connecting point for intermodal transportation systems (Jeevan et al., 2015). Table 4.2 depicts the major components of the Malaysian inland ports consisting of container throughput capacities and main features of inland port terminals.

A container terminal's capacity is usually measured in terms of the number of containers handled by the terminal per year. Since containers come in two standard sizes that are 20 feet and 40 feet, the capacity is usually expressed in 20-ft, container equivalents called TEU. An essential characteristic of containerisation is the space requirements for the terminals. The storage of containers in an elevated depot, or on remote inland sites pushes up operating costs in the port. Therefore, good planning of the area requirements must be done before the development of the port.

The rail network has the main industrial catchment areas such as Ipoh and Klang Valley. It connects other transport nodes such as major container seaports (Penang Port, Port Klang and PTP) and four inland ports (located near Padang Besar, Ipoh, Seremban and Segamat). The three container seaports have been provided with a train service of six trips per day connecting to other hinterland nodes. Each train service can carry 64 TEUs per trip, which is lower than the world average capacity of 66 TEUs per trip (Woodburn, 2011). The double-track railway network has increased to 774 km (47%) after completing the electrified double-tracking project from Ipoh to Padang Besar in November 2014 (ASEAN-Japan Transport Partnership, 2015). However, railway lines use metre-gauge (narrow gauge) will limit the ability to provide the double-stack container service (Malaysia Freight Transport, 2012).

Table 4.2 Malaysian inland ports features and capacities

Inland ports	Features	Capacity
Padang Besar Cargo Terminal (PBCT)	PBCT is the main gateway for southern Thailand. Railway station at Padang Besar will expand to accommodate more cargo from Thailand, which would then head for onward shipment via Penang Port	Cargo handling capacity 120,000 TEUs annually to over 150,000 TEUs
Ipoh Cargo Terminal (ICT)	Facilitates import-export for Kinta Valley businesses, distribute imports from Port Klang, Penang Port and Johor port	Container Yard No 1: Laden/Empty Stacking 1.8 ha (3.96 acres) Capacity 800 TEUs—3 high Container Yard No 2: Laden/Empty Stacking 3.0 ha (6.60 acres). Capacity 1440 TEUs—4 high
Segamat Inland Port (SIP)	Located in Segamat, Northern Johor for the surrounding industrial parks operating as a centre for the transhipment of sea cargo to inland destinations. SIP extends its container storage depot facilities in Segamat and Pasir Gudang. It offers feeder rail links and roads to Port Klang to shippers in the South-central region to various seaports. Including facilities: inland port terminals, port handling machines, railway siding terminals, customs office, container depot, haulage depot, bonded warehouses, open storage yard, weighbridge, commercial offices, etc.	Cargo handling capacity: Capacity 10,000 TEUs. Both of the container depots have a combined storage capacity of up to 10,000 TEUs. The container depots provide storage for dry boxes, ISO tanks and reefers
Nilai Inland Port (NIP)	NIP is an intermodal container hub for Port Klang and Port of Tanjung Pelepas (PTP). It provides road/rail links and logistics services to industries in the Central Region	Capacity 24,000 TEUs

(continued)

Table 4.2 (continued)

Inland ports	Features	Capacity
Tebedu Inland Port (TIP)	TIP provides a link between the Malaysian state of Sarawak and Indonesia's Kalimantan region on Borneo's island. TIP offers all the necessary facilities and services such as round-the-clock security, open yard, container yard, warehouses, transfer sheds, a fleet of forklifts and labour. Loading and unloading cargo are carried out in an orderly manner amidst a safe and conducive area	Facilities—Storage: Warehouse 23,000 sq ft Container Yard: 196,020 sq. ft Refrigerated Storage—Free Storage: Full container—3 calendar days Empty container—7 calendar days

Intermodal transportation requires improvements in the efficiency of transport services. The logistics performance index (LPI), which indicates the quality of trades, transport and services, varies widely among Asian countries. For example, Asian countries with a high LPI for 2016 include Singapore (4.14), Japan (3.97), Republic Korea (3.72), Malaysia (3.43) and Indonesia (2.98). Most landlocked countries have low LPI—road transport, which is the most flexible transport mode. Usually operated and managed by the private sectors, and in certain limited cases, by public sector subsidiaries. The public sector railway companies still handle most railway route operations in Asia under overall government control, especially in Malaysia. Quality is the single most significant problem in railway freight operations.

Nevertheless, it has been discovered that the capacities of Malaysian inland ports are underutilised; thus, their expected benefits to relevant stakeholders appear limited (Jeevan et al., 2015). In order to improve the capacity, utilisation and operational efficiency of Malaysian inland ports, it is crucial to identify the challenges facing them and factors influencing their operations. Infrastructure and transport facilities such as good highways and wide roads and sufficient operational equipment are the major concerns for Malaysian inland port operations in terms of capacity utilisation. The road system facilitates almost 80% of freight transport (PWD, 2014). For instance, most containers to and from significant seaports used a single transportation mode, i.e. road. Without other modes of transportation (for example, connecting from Nilai Inland Port to Port Klang and Port of Tanjung Pelepas), the overutilisation of road infrastructure has dramatically affected roads' condition in this region. This creates congestion on the roads and highway, which results in delays in shipping schedule to seaports, affecting shippers' costs and time. For example, in Port Klang, road lanes' existing narrow width introduces instability to freight transportation, creating significant overloading pressure and reducing road conditions.

Unfortunately, not all Malaysian inland ports, including seaports, are located strategically to assist their stakeholders. A high percentage (70%) of the expert reviewers agreed that the Northern Corridor Economic Region (NCER) has a strong potential for successful intermodal transportation (Nasir, 2014). Robust development of the industrial area for NCER, particularly in Bukit Kayu Hitam has made this corridor a potential location. In line with One Belt One Road (OBOR) initiative (Strategic land Bridge Hub), the northern region of peninsular Malaysia (proposed in the area of Kulim and Bukit Kayu Hitam) can serve as the main gateway to the south of the Thailand market. Bukit Kayu Hitam can serve as a new transhipment hub. However, the corridor needs a new rail connection and inland terminal for intermodal transportation in this region (Nasir, 2014). These features will make the existence of an inland port in the northern region of peninsular Malaysia to be logistically relevant in holding a strong foothold for a successful intermodal transportation hub in the Northern Corridor Economic Region (NCER). As the other alternative to the south of Thailand (which is located in Padang Besar of the state of Perlis) is already highly congested. Padang Besar-Johor Bahru rail is part of the Singapore-Kunming Rail Link network.

4.3 Conclusion

Inland ports replicate the services and facilities of seaports. For example, Singapore Port in 2015, due to its comprehensiveness of port facilities and services were unsurpassed in terms of quality, efficiency, competitiveness and reliability; with its ability to handle over 2,000 containers per vessel, with a turnaround time of fewer than 12 h. The Asian Freight and Supply Chain Awards (AFSCA) have frequently recognised Singapore as the 'Best Seaport in Asia'. Singapore was also triumphed by other major ports in the region - Hong Kong (74), Shenzhen (133) and Malaysia's Port Klang (69) according to the OECD ITF report. Singapore's container port moved an average of 66 containers per hour per ship in 2014. Its geographically closest rival, the Port of Tanjung Pelepas in southwestern Johor in Malaysia, racked up an average of 81. Nevertheless, Shanghai port chalked up 167 container moves per hour per ship, the world's highest in 2014.

The factor that affects an inland port's performance is the turnaround time of the inland port and the nearest seaport. In general, turnaround time includes the time spent on entering the port, loading, unloading, and departing. The factors that affect the ports turnaround time are berth times and delays of container ships, dwell time, container cargo and truck turnaround time, and customs clearance duration. The delay in an inland port turnaround time might be due to inadequate quay/gantry crane equipment, limited storage capacity, and poor multimodal connections to hinterland and infrastructure. Even though turnaround time is relevant for shipping lines (related port costs), ports do not typically report turnaround times in their annual reports or other publications.

Identification and implementation of all the factors that affect the inland port's performance will gain a few advantages. They are,

- Achieve efficiency and sustainable options through rail transportation and multimodal.
- Increase the capacity of freight distribution through multimodal, intermodal and supply chain system efficiency.
- Consolidate import, export and distribution activities in one location.
- Increase the availability of space for warehouse and distribution centres.
- Improve speed to market with the lowest cost of shipping provided a rail link is available.

Based on the reviews of an inland port's essential logistical factors, it is therefore concluded that an inland port should play a more strategic and creative role in managing the demands and planning to make it an important distribution hub. This is especially true for the northern region of peninsular Malaysia (proposed in the area of Kulim and Bukit Kayu Hitam) in line with One Belt One Road (OBOR) initiative (Strategic Land Bridge Hub).

References

Arvis, J. F., Saslavsky, D., Ojala, L., Shepherd, B., Busch, C., & Raj, A. (2014). *Connecting to compete 2014: Trade logistics in the global economy—The logistics performance index and its indicators.*

ASEAN-Japan Transport Partnership. (2015). ASEAN—Japan Transport Partnership. Retrieved January 12, 2018, from http://www.ajtpweb.org/statistics/Malaysia/MalaysiaAJTP2014/rail-transport.

Beresford, A., Pettit, S., Xu, Q., & Williams, S. (2012). A study of dry port development in China. *Maritime Economics & Logistics, 14*(1), 73–98.

Cho, H. S. (2014). Determinants and effects of logistics costs in container ports: The transaction cost economics perspective. *The Asian Journal of Shipping and Logistics, 30*(2), 193–215.

Christiaanse, E., & Kumar, K. (2000). ICT enabled coordination dynamics supply webs. *Journal of Physical Distribution and Logistics Management, 30,* 268–286.

Hanaoka, S., & Regmi, M. B. (2011). Promoting intermodal freight transport through the development of dry ports in Asia: An environmental perspective. *IATSS Research, 35*(1), 16–23.

Jeevan, J., Chen, S. L., & Lee, E. S. (2015). The challenges of Malaysian dry ports development. *The Asian Journal of Shipping and Logistics, 31*(1), 109–134.

Khalid, N. (2012, December 01). *Measuring Performance of Malaysian Container Ports.* Center for Maritime Economics and Industries Maritime Institute of Malaysia.

Malaysia Freight Transport. (2012). *Part of BMI's industry survey and forecasts series.* 85 Queen Victoria Street London.

Monios, J. (2011). The role of inland terminal development in the hinterland access strategies of Spanish ports. *Research in Transportation Economics, 33*(1), 59–66.

Nasir, S. (2014). *Intermodal container transport logistics to and from Malaysian ports: Evaluation of customer requirements and environmental effects* (Doctoral dissertation, KTH Royal Institute of Technology).

PWD. (2014). *Public Work Department.* Malaysian Highway Authority. Retrieved September 16, 2018, from http://www.kkr.gov.my/ms/node/118.

Rodrigue, J. P., Debrie, J., Fremont, A., & Gouvernal, E. (2010). Functions and actors of inland ports: European and North American dynamics. *Journal of Transport Geography, 18*(4), 519–529.

Roso, V., & Lumsden, K. (2009). The dry port concept: Moving seaport activities inland. *UNESCAP, Transport and Communications Bulletin for Asia and the Pacific, 5*(78), 87–102.

UNCTAD. (1991). *World Investment Report 1991* [Online]. Retrieve on February 20, 2018, https://unctad.org/webflyer/world-investment-report-1991.

UNECE. (2009a). *Illustrated Glossary for Transport Statistics* [Online]. Retrieved February 20, 2018, from http://www.unece.org/fileamin/DAM/trans/main/wp6/pdf.

UNECE. (2009b). *UNECE Division Database, compiled from national and international (EUROSTAT, OECD, CIS) official sources.* Retrieved February 15, 2018, from http://w3.unece.org/pxweb/Dialog/Saveshow.asp. Consulted 17.5.2009.

Woodburn, A. (2011). An investigation of container train service provision and load factors in Great Britain. *EJTIR, 2*(11).

Ts Dr. Adam Mohd Saifudin is an Associate Professor in School of Technology Management and Logistics (STML) and obtained his Doctor of Business Administration (DBA) in 2012, UUM, specialised in Logistics Warehouse Operations and Management. His research expertise is in Warehouse, Logistics Management, and Halal Supply Chain and has published in journals and presented many papers in the international conferences. He is a Fellow and Vice President of The Society of Logisticians Malaysia (LogM), Professional Specialist (Logistics and Transportation) and Chartered Member of Chartered Institute of Logistics and Transportation (CMILT), UK. Before joining UUM as an academician, Dr. Adam was with the industries for 20 years and started as a Journalist with a major national newspaper in 1986. He then worked with a local company dealing with oil and gas as Regional Manager in Penang before entering into manufacturing with Sony Electronics (M) Sdn Bhd (SEM) in 1992. He gained many successes during his tenure as senior staff and specialised in Material Operations and Environmental Management System (EMS). As one of the key senior staff of the EMS/ISO 14000 team, he coordinated and managing the major success of the company by attaining the first certification for the company in the country in 1996 and later guided few other sister companies—Sony Mobile Electronics Thailand (SMET); Sony Electronics Indonesia (SEI); and Sony Audio Product Malaysia (SAPM) to attain the certification the following year.

Nizamuddin Zainuddin is the Senior Lecturer and Coordinator for Maritime Management course. He holds a Master's Degree in Business Administration from Keele University, UK and an Honours Bachelor Degree in Business Administration (Transportation Management) UiTM, Shah Alam, Malaysia. He is also a Chartered Member of Chartered Institute of Logistics and Transportation (CMILT), UK; Member of The Society of Logisticians Malaysia (MLogM); Fellow Researcher for the Center of Logistics and Transportation (CeLT); Fellow Researcher for the Technology Supply Chain Center (TeSCe); and Fellow Researcher for the Disaster Management Institute (DMI) UUM. He has researched and published journals and conference papers on the logistics supply chain, halal logistics, inland port development, sustainable transport, humanitarian logistics and road safety management. He has also supervised many undergrad and postgrad students. His research interests include logistics management, halal logistics, halal supply chain, maritime management, humanitarian logistics, road safety management and inland port development.

Chapter 5
Think Global, Act Local: Strategies to Develop a Sustainable Mobility Programme in Promoting Socio-Economic Development for a Better Quality Education

Farah Azea Khalid and Yazrina Yahya

Abstract Sustaining a mobility programme at higher learning institutions requires dedication and commitment from all parties involved. However, it is difficult to sustain a mobility programme due to several factors such as financial support, marketability of the programme, the number of participants attending, and the institution's political changes. In view of this, a good reference model is required to identify the challenges and provide the solution needed to sustain the mobility programme. A model has been developed, and it provides the identification of challenges and detailed solutions required. The model's success is discussed through its implementation on a mobility programme known as UKM-GSMP, which has been running for more than a decade. This had further proved that with proper model, strategy and execution, it is possible to sustain a mobility programme for a period of time. Stakeholders' interviews showed that the program has provided the students with good experience, and knowledge and helped them understand the local culture, leading to better education quality. The programme had also benefited the institution socio-economic development via the income generation and good branding for the university.

F. A. Khalid
Faculty of Social Sciences and Humanities, Universiti Kebangsaan Malaysia,
Bandar Baru Bangi, Malaysia
e-mail: farahazea@gmail.com

Y. Yahya (✉)
Faculty of Economics and Management, Universiti Kebangsaan Malaysia,
Bandar Baru Bangi, Malaysia
e-mail: yazrina@ukm.edu.my

N. F. Elias et al. (eds.), *Service Excellence for Sustainability*,
https://doi.org/10.1007/978-981-16-2579-4_5

5.1 Introduction

The purpose of education, as put forth by John Dewey (1916), is to equip every student with knowledge and experience in preparation for them to enter post-graduation life as engaged and informed citizens. Universities are the platform that provides educational experiences that challenge and prep students to serve as contributing members of diverse communities. Taking into consideration the context and complex issues present around the world in the twenty-first century, universities should educate, prepare, and empower students to resolve problems contemporary in the society by going for the development of global citizenship approach.

The higher-ups of the higher education administration constantly encourage students to study abroad, setting ambitious percentage goals in the name of "Internationalisation of higher education." There are a variety of opinions as to why this radical transformation is taking place—but most seem to agree on the notion of globalisation (Lewin, 2009). To be up-to-date with this continual growth, universities today are putting effort in establishing platforms for students to go abroad in addition to practising internationalisation in home universities. The majority of universities today has developed Student Mobility and Study Abroad programmes such as exchange, summer, and specialised tailored needs as a forthcoming step towards globalisation.

Student Mobility, by definition from the University of Glasgow (2016), is "any opportunity for students to work or study abroad whilst undertaking their degree programme—whether undergraduate or postgraduate." Student mobility can be of inbound and outbound with partner institutions bilateral exchange collaboration or one-way study abroad for durations that vary from 1 week to a year. Student mobility opportunities are believed to be an integral part of an academic programme to enhance the overall student experience in education, extra-curricular, and socialisation. It helps broaden their horizons as they get to experience a different academic environment.

The concept of global citizenship is that human beings are "citizens of the world" (Dower & Williams, 2002). A good global citizen is a contributive member of the wider community of humanity and the world. Implementing globalisation and internationalisation in varsity students is formulated to serve this purpose. Despite the many benefits and successes of mobility programmes, the key problem faced by Universiti Kebangsaan Malaysia is in sustaining these programmes. Countless mobility programmes have been introduced, however, not many have reached the bar of 10 years' sustainability. Questions raised are on how these mobility programmes can be sustainable; and why they need to be. In this paper, we delve deeper into the insights of these questions in the following sections.

5.2 Development of a Sustainable Mobility Programme

Globalisation is deemed as an important concept in understanding the world. The term "think global, act local" was coined by Stuart Grauer in his 1989 University of San Diego publication, and has since been applied to international education initiatives. Therefore, understanding the concept of "think global, act local" should be implemented in varsity students as they are fundamental to our future (Divitni et al., 2011). To guarantee the development of sustainable mobility programmes, it is crucial to weigh on what the global community wants.

Students going for inbound and outbound programmes seek local attractions such as low-living costs, multiculturalism, quality education, and employment prospects. Based on these factors, universities can design mobility programmes that meet the demands of the global community. Elements of a sustainable programme have been mentioned in Muhammad Faizul et al. (2018) "The Good, The Bad, and The Ugly" that discusses the issue at hand of sustainable mobility programmes for a period of time, and suggestions on how to overcome the challenges.

The "Model for Sustainable Mobility Programme" that was introduced in the article is divided into three subcategories namely "The Good", "The Bad/Ugly: Challenges", and "Sustainable Solutions". The challenges that were put forth were troubles in sustaining, lack of participation, financial funding, social adaptation, marketability and marketing strategies, a variation of the global economy, ecology changes, and political stability. In order to overcome these challenges, Muhammad Faizul et al. (2018) suggested four solutions. Firstly, universities and institutions should secure the support of stakeholders and partners. Secondly, students' quality must be guaranteed by ensuring them to be all-rounders with unique identities to represent their home universities as they carry the varsity and country's good name.

Thirdly, universities should take initiatives in product offering and marketing. Based on the feedback from participants of inbound mobility programmes of UKM, the majority claims that the cultural learning experience alongside a diverse group of students outside class hours has left a strong impression of the university and Malaysia. This is also the basis of the fourth solution, highlighting the importance of cultural activities and Student Buddy strategy. UKM's International Relations Centre (UKM Global) is strongly supported by a team of Student Buddies represented by participants of previous mobility programmes and volunteers who are passionate students eager to be a part of internationalisation in the university. Student Buddies carry multiple responsibilities ranging from assisting inbound students during the registration period to making them feel at home in a foreign land by offering genuine friendship.

Although the model by Muhammad Faizul et al. (2018) has covered substantial details and solutions to assist in developing a sustainable mobility programme, there is still a dilemma associated with the model's design. Proper composition of the preliminary model is required and crucial for a better understanding to further execute the model when used to design mobility programmes.

5.3 Revised: Elements of a Successful Sustainable Mobility Programme

As previously raised, further investigation is performed to determine a better model. The existing model is expanded by carefully examining all the elements mentioned above, going into detail of the actions and effects of implementing this model; in order to achieve the goal of a successful sustainable programme. The investigation results reveal that the existing model provides physical components and solutions, excluding further details and refinements. Although the components are mentioned, its inclusion and further details would provide better understanding and give a better description of how every component can be implemented.

The newly expanded model is enhanced on the sustainable solutions component. The existing model put forth solutions such as support from stakeholders and partners, good quality of students, programmes: cultural activities and student buddies; and product offering and marketing. These solutions are further elaborated and refined to provide a better composition in giving the process involved for executing each solution. The differences between each model are shown in Table 5.1.

In providing further details of each solution, an example based on a case study at the International Relations Centre UKM is given to assist the execution of each solution. The new solutions identified for the expanded model is explained in the following sections.

Table 5.1 Comparison of existing models and enhanced model

Existing model	Enhanced model
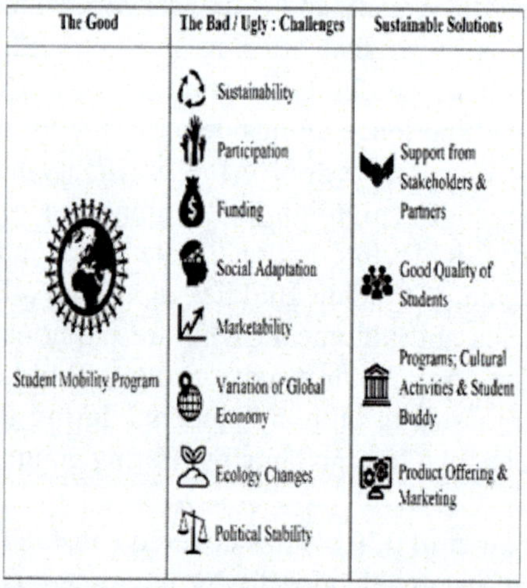	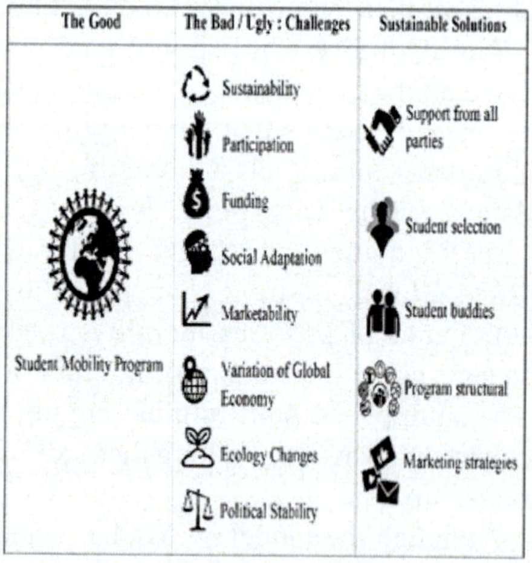

5.4 Support from All Parties

Muhammad Faizul et al. (2018) deliberated the component support from stakeholders and partners, focusing on the definition and stakeholders' types. Hence, the new solutions provided focus on the kinds of support given by/to the stakeholders and partners. Support from all parties refers to the process involved in getting to know the partners, finding mutual understanding, and collaborating with them.

In the endeavour to perpetuate a mobility programme, university or organisation should ensure the support from all parties involved. Before collaborating with higher education institutions, we should be familiar with our potential partner(s). Getting to know them in detail is advisable before investing in a possible foundation for future partnerships. Doing a background check on their visions, strategies, academic expertise, and past international collaborations helps ensure that their institution is on the same page as ours by critically comparing these elements with our profile. If their ambitions are compatible with ours, there will be a shared understanding of effective and constructive expectations. By developing a shared vision, this would encourage long-term partnerships.

With mutual understanding and a shared vision, partner universities need institutional commitment to maintain a good relationship. Commitment from all parties can be enthused by multiple methods such as formalising the relationship, tasks, and responsibilities of administration and integrating them into the institutional policy; and continuously monitoring the partnership using both quantitative and qualitative indicators. When there is a strong synergy between partners, it ensures longevity. The factors that are deemed essential in creating sustainable partnerships are continuous compatibility and long-term commitment that is institutionalised at all parties' levels.

Regular exchange of students and collaborative academic activities can indefinitely contribute to ensuring partnerships' sustainability, thus leading to ceaseless mobility programmes. Following this, well-established partnerships are more welcoming for students to participate; which leads to our next element of student selection.

5.5 Student Selection

Student selection refers to students chosen to go for mobility programmes; who should be of quality as they represent the varsity and country. To select suitable participants for these programmes, we have to set a benchmark for candidates who wish to participate. What defines a perfect student? A top scorer? Rather than people who want a grade, it is better to get people who want to learn. Of course, academic performance has to be excellent as well; hence setting minimum academic requirements is necessary. Besides, student of quality must be disciplined and possess critical intelligence. A good student knows how to balance academics with extra-curricular activities to achieve the target of overall personality development. By making sure quality participants, this allows them to engage in intellectual discussions with critical mindsets.

At home universities, some preparations should be made to ensure the quality of students. We need to prep them up before they go for mobility programmes by gathering the chosen candidates and thoroughly explaining their responsibilities and expectations before, during, and after participating in the mobility programmes. This method allows them to plan ahead and always remain in pursuit of knowledge.

5.6 Student Buddies

Aside from getting the students to go abroad, there is a need to have the student buddies at the home or host universities. The buddies are important to provide additional support and giving good student experiences to the international inbound students. On the other hand, being a student buddy is a golden opportunity for undergraduates to explore their identities and build their leadership and citizenship skills. A good selection of student buddies is necessary for ensuring their commitment and involvement, following related training for that matter. Hence in UKM, the International Relations Centre provides a training course at the beginning of every semester where past participants of mobility programmes are gathered upon their return. The purpose of the training course is to bring them together, developing their leadership skills, and present the benefits of being committed to the team. Moreover, this offers experiential learning opportunities to engage with diverse students, encouraging what we refer as our Student Ambassadors to explore their positions in our increasingly international societies. UKM Global also strongly welcomes students who have never participated in mobility programmes but are genuinely interested in joining the team.

UKM Global's Student Ambassadors function as an intermediary between the office and international students who come here for mobility programmes, after a thorough training on hospitability and intercultural sensitivity. As Student Ambassadors have to engage with people from different parts of the world, UKM Global sterns individuals to respectfully interact with different cultures and make them feel like home in this foreign land by assigning them to take care of the students from different countries across the globe. It is a great platform for them to apply their influence, skills, and knowledge to practice leadership and good citizenship that advance positive social change in society. Student Ambassadors carry the responsibility of displaying a good image of UKM and Malaysians.

5.7 Programme Structure

Another element of a successful sustainable mobility programme is possessing a well-structured composition. Cooperation between the International Relations Centre with faculties and coordinators of every party is crucial in assuring the programme's smooth-sailing. Disclosure helps in avoiding miscommunications that might cause inconvenience to the participants. In addition, we should establish anchors and champions for more structured reliability among partnerships.

Furthermore, the inclusion of cultural components and offering niche programmes would appeal to the global community and ceaseless welcome demands of participation. This is because participants from across the world wish to explore the richness of Malaysian culture. Hence offering niche programmes unavailable in other countries such as tropical heritage, indigenous communities, Islamic banking, and Syariah law would arouse potential participants' interest to learn more about what we have to offer.

Based on the successes of mobility programmes in UKM, these contributing factors have proven effective. Further evidence is provided in a case study in Sect. 5.10.

5.8 Marketing Strategy

Marketing strategy is very important as a mobility programme cannot be sustained without the aid of a proper marketing strategy, despite having a well-structured programme. Marketing strategy is the comprehensive plan formulated particularly for achieving the organisation's marketing objectives, as stated by Juneja (2018). In the context of varsity mobility programmes, the objectives are to generate income by increasing partnerships and participation. Hence to achieve this goal, it is crucial for us to identify a targeted market.

It is important for the university to identify their strategic partners and regions based on the university's needs. Factors such as available research and mobility grants, activeness in sending and receiving international students, an offering of good summer and other exchange programmes must be taken into consideration. Some of UKM's distinguished targeted markets are the New Colombo Plan (NCP) of Australia, and Grants and Funding of Japan; where funds are made available. Outreach via conferences of The Asia-Pacific Association for International Education (APAIE), European Association for International Education (EAIE), and Quacquarelli Symonds (QS) on the other hand provides the platform to obtain huge numbers of partners in within a short period.

Besides, social media strategy is equivalently important to keep up with the trends of the twenty-first century to intrigue the younger generation. This can be supported by contributions from students or past participants in the form of videos or blog entries of their exciting experiences to further market the mobility programme to future participants. The university is also responsible for constantly updating social media with strong visuals to alert students of the opportunities we offer. Last but not least, seeking media organisations such as TV and radio programmes for publicity helps in marketing strategies as well.

5.9 Why a Programme Need to Be Sustainable?

The elements that lead to a sustainable mobility programme have been identified, and it is essential to know the importance of sustaining the programme. A sustainable programme will certainly promote the local students' socio economic development and assist in providing better quality education. A sustainable programme for more than ten years will provide a good platform for local students to experience internationalisation at home consistently. It will also provide stable socioeconomic support for the university as the fees received is a source of income to the university.

Also, it is beneficial for the internationalisation of our home universities. Internationalisation of higher education is a means of improving education quality, not only as a goal in itself (Jibeen & Khan, 2015). Internationalisation in higher education carries multiple benefits in sustaining dynamic academic exchanges and building partner countries' social and economic capacity.

Moreover, it helps materialise the concept of "think global, act local". This develops the global community at home universities. By giving them proper exposure and opportunity to interact with people from different cultural backgrounds, they also share the richness of Malaysian culture. This trains our students to have a global mindset but rich with local culture, simultaneously increasing their marketability in the working world.

Where there is stability, there will be higher demands for collaborations. Creating a sustainable mobility programme can also generate income for the university. In addition to this, it contributes to the good image niche branding for the university. When the university possesses a notable reputation, there will be continuous supply of participants eager to be part of the experience. Hence, it will create a balance in the academic ecosystem based on our product offering. The efficiency of the elements mentioned above is further supported by evidence from our case study.

5.10 Case Study: UKM-GSMP

The mobility programme UKM-GSMP has been running for more than ten years, making it a successfully sustainable programme that has produced a huge number of alumni. The success of the programme is translated into the accomplishment of support from all partners, student selection, student buddies, programme structure and its marketing strategy. Initiated a decade ago, UKM-GSMP continues to be the strongest platform for the reformation of higher education and internationalisation in UKM, at par with its slogan "Building Bridges for International Understanding and Friendship". The 10th UKM-GSMP reflects the continuous, ascending success of this programme.

Since a decade ago, students from over the world have participated in the UKM-Global Student Mobility Partnership programme. Initiated 10 years ago, the UKM-GSMP has produced a total of 387 alumni to date. In 2018, 22 students from Mexico, five from Australia, and three from Hong Kong were chosen alongside five UKM's local students for this 3-weeks summer mobility programme. The participants were divided into two courses, namely Indigenous Communities in Malaysia (ICM) and Sustainability of Tropical Heritage (STH). They went for field-trips to the relevant sites in Peninsular Malaysia, such as Taman Negara for ICM and Langkawi UNESCO Global Geopark for STH. The participants were prepped with theoretical and classroom-based lectures during their first week before embarking their challenging journey for the fortnight to come.

This programme running for a decade, further proves the efficiency of a sustainable mobility programme's previously mentioned elements. "I chose to participate in this programme because I wanted to learn about environmental issues and ways to overcome them. However, alongside getting that knowledge by the end of this programme, I have also gained much more than just that," Modesto, a mass-communication student from New Mexico University, said. Local and foreign participants were interviewed to get their constructive feedback.

When asked about what they have learned throughout the programme, most of them claimed that it had been an enriching experience that broadens their knowledge about Malaysia, and the world in general. Malaysia, possessing an ecosystem quite different than that of Hong Kong, Mexico, and Australia, has taught them to be tolerable and respectful of different cultures. In addition to this, our local participants added that they learned things about Malaysia that they were never aware of, such as our indigenous communities' beauty and issues. The firsthand experience they had gained from living with the community had been eye-opening and having the opportunity to discover all that alongside people from various nationalities induced a strong appreciation for Malaysia's rich cultural diversity.

This programme also reflected the student buddy programme's effectiveness as the five local participants were given former training on hospitability. When asked about the most memorable part of the programme, Chava from Mexico highlighted how "Malaysians are really friendly and embraced him despite his nationality." The other participants had mentioned on the interesting places that they were taken to throughout their stay. However, the significant similarity in their answers was that they mentioned how the people they were with were the best part of it.

Even though they had to face inevitable hardships, all of them agreed that they would definitely recommend this programme to other people in their home countries. Susan from Mexico, who was invited to our local TV show *Selamat Pagi Malaysia* enthusiastically said, "Of course. Going for an exchange gives you lots of experience and broaden your views. You're in a foreign place, but you'll feel like you're at home with a new family." Modesto added on during his interview on Traxx FM, "It gives you a lot of knowledge. Not only theoretical and classroom-based, but also firsthand experience of different cultures from people all over the world – it makes you grow in just three weeks."

Nelson, one of UKM Global Student Ambassador and also a lucky participant of the 10th GSMP further said, "Initially, I didn't think that I'd be so involved as a participant along with the other international students because I thought I would just be a student buddy to aid and help facilitate the programme. However, I got to join the activities and learn countless of things at the same time. I'm so honoured and thankful to be given this opportunity to be a part of this year's GSMP which happens to be the 10th anniversary. I couldn't have asked for a better summer, and this will always be one of the highlights of my university life."

"I will be 15,000 km away, but Malaysia and all of you will always be in my heart. I will miss all of you, and lots of love," Lizeth, from Mexico, represented her peers during the closing ceremony of the 10th UKM-GSMP and delivered a heart-warming speech. It summed up their endless gratitude towards both sides' coordinators for the knowledge, opportunities, and hospitality they received during this life-changing experience.

Participants from previous GSMP were also contacted to determine whether this programme has left an imprint in their lives. Hidayah Noar, a participant of the 8th GSMP answered, "GSMP will always have a special place in my heart – an adventure I would never trade for another. The programme has definitely broadened my horizon and opened up doors to many other opportunities. Most importantly, I will forever cherish the life-long friendship that was built over those three weeks. GSMP truly lives up to its tagline – building bridges for international understanding and friendship."

Ivan Ong, a participant of the 9th GSMP claimed that GSMP is indeed a once in a lifetime experience. "I thank UKM Global for this wonderful opportunity. I got to be friends with people from the other side of the world, coming from different cultures and beliefs, yet we became so close to each other. It has been over a year since then, but we still keep in touch. The network that I have created makes me feel blessed to be part of the experience."

To commemorate their precious, unforgettable experiences, UKM Global has created a Facebook page specifically for GSMP. It has been running since the first GSMP until the tenth, treasuring all the memories for them to look back at and inspiring other people who are intrigued to join this summer mobility programme. It is also a means of communication for the past participants to keep in touch with one another and forward inquiries to UKM Global. This further proves the efficiency of being adept in social media strategy.

5.11 Conclusion—The Way Forward

Sustaining a mobility programme for more than a decade requires dedication and commitment from all parties. However, without a proper model and strategy, it will be challenging to do so as executors or coordinators of the programmes come and go. People who are put in charge of the programme may not be the same people over the years; hence a good reference model is required.

The model proposed has provided the details on how a programme can be sustained. By carefully following the solutions given, a programme will be able to be sustained for a longer period of time. This has been proven by the case study mentioned; UKM-GSMP.

UKM-GSMP is genuinely the best example of a successful sustainable mobility programme because of the diligent actions taken in ensuring so. The programme is sustainable because of the structural competence where all parties' coordinators are dedicated, content is appropriately planned and executed, with strong continuous support from stakeholders. As the programme grows over the years, it leads to a compelling branding and marketing strength for the university. The participants' quality was also well taken care of as only intellectual students looking for a global citizenship change were handpicked to be a part of this experience. Last but not least, the success of this programme can also be credited to the dedication of student buddies.

Sustaining a mobility programme for more than a decade requires hard work and dedication. Without proper strategy, model, and execution, it would not be easy. Based on our case study, the students' feedback reflects the benefits they garnered from the programme and how the alumni contribute to the university's ecosystem. Almost every stakeholder involved has gained not only good experience, knowledge, good branding image, but also the global community with the understanding of the local culture. There are too many benefits to sustaining a good mobility programme. Hence, universities should step up and take action in doing so.

References

Dewey, J. (1916). *Democracy and education: An introduction to the philosophy education.* New York: Macmillan.

Divitni, M., Jenssen, M., & Skau, K. (2011). *Think globally, act locally: Promoting pre-school learning about other cultures.* Norwegian: Department of Information and Computer Science, Norwegian University of Science and Technology.

Dower, N., & Williams, J. (2002). *Global citizenship: A critical introduction.* New York: Routledge.

Jibeen, T., & Khan, M. A. (2015). Internationalisation of higher education: Potential benefits and costs. *International Journal of Evaluation and Research in Education (IJERE), 4*(4), 196–199.

Juneja, P. (2018). *Marketing strategy—Meaning and its importance.* Management Study Guide. https://www.managementstudyguide.com/marketing-strategy.htm [30 September 2018].

Lewin, R. (2009). *The handbook of practice and research in study abroad: Higher education and the quest for global citizenship.* New York: Routledge.

Muhammad Faizul, A. H., Yahya, Y., & Ahmad, A. L. (2018). The good, the bad and the ugly: A model for sustainable mobility programme. In N. Mohd Nordin, Y. Yahya, M. Mustafa, & N. Sarmin (Eds.), *A chapter in a book "Internationalization and student mobility for higher education institutions in Malaysia"*. USM Publisher.
University of Glasgow. (2016). *Student mobility*. https://www.gla.ac.uk/media/media_242620_en.pdf [9 October 2018].

Farah Azea Khalid is a graduate from Universiti Kebangsaan Malaysia. She holds a Bachelor Degree in English Language studies. She is an experienced student buddy who has been assisting international students at the International Relations Centre, UKM. She has been abroad as a student exchange and very well verse in the intercultural communication programs.

Dr. Yazrina Yahya is the Deputy Director of Higher Education Leader-ship Academy (AKEPT), Malaysian Ministry of Higher Education. She was previously the Director of Universiti Kebangsaan Malaysia (UKM) Global. She is recognized for her expertise in the area of Internationalization of Higher Education and Information Technology where she has been invited as a keynote speaker by Universiti Kebangsaan Malaysia; Universitas Gadjah Mada, Bulacan State University, University of Min-danao, Philippines; Universiti Andalas, Politeknik Padang, Indonesia, Commission of Higher Education (CHED) Philippines and Cebu Normal University Philippines. She was also invited as a guest speaker by the Canadian Bureau of International Education (CBIE), EU-SHARE, ASEAN University Network (AUN), Sophia University, Japan, Universiti Andalas, Australian High Commission in Malaysia, and CAMPUS France in various international conferences held globally and locally. She was also one of the key persons to head the collaboration between the Department of Education Mexico (EDOMEX) and UKM, resulting in UKM hosting 100 Mexican students for a three-week summer program at various faculties in 2013. Partners from the ASEAN University Network (AUN), Association of University of Asia Pacific, EU-SHARE and ERASMUS Techno II have also acknowledged her expertise in managing international forums and securing student numbers to go abroad for exchange.

Chapter 6
Media Psychology in New Era Communication

Huzili Hussin, Adila Ismail, and Mohammad Rezal Hamzah

Abstract Over a decade ago, media psychology emerged as a branch of discipline in psychology which studies the interaction between media technology and human beings. Nowadays, the focus of study has a shift to look at the effects of media technology on a human being. The discussion had made a significant remark to discuss what technology had done to our society critically. Psychology is the key to understand the implications of technology. Media psychologists' goal is to find the answers and solutions by combining an understanding of human behavior, cognition, and emotions. However, media psychology is not just concerned with content, but it looks at the whole system. Just as the reciprocal relationship between environment, behavior, and cognition, media psychology evaluates the system's interactive process. The rapid development of technology has triggered a variety of reactions, from enthusiasm to distrust. As technology changes our lives, we are forced to change how we view the world. The key to media psychology is you learn to understand psychology and technology. The tools of media psychology help individuals to take responsibility and part of the system. It is the only way to develop better technologies and use them well.

6.1 Introduction

The subject matter of media psychology has developed as a distinct subdivision of psychology within the last 25–30 years, even though it has been around for decades. In Hugo Münsterberg's early work on psychology and the film's psychological impact,

---

H. Hussin (✉) · A. Ismail · M. R. Hamzah
Centre of Excellence for Social Innovation and Sustainability, Faculty of Applied and Human Sciences, Universiti Malaysia Perlis, Arau, Malaysia
e-mail: huzili@unimap.edu.my

A. Ismail
e-mail: adilaismail@unimap.edu.my

M. R. Hamzah
e-mail: rezal@unimap.edu.my

N. F. Elias et al. (eds.), *Service Excellence for Sustainability*,
https://doi.org/10.1007/978-981-16-2579-4_6

media psychology found its inspirational origins more than 100 years ago (Fischoff, 2005). Published as *The Photoplay: A Psychological Study*, Münsterberg (1916) was the first empirical study of audience feedback to a film and his work is still being referenced today in numerous film, schools, and seminars since it provided such a thorough analysis of visual architecture and cinematographic norms and their psychological impact on the public. This shows that studies on the effects of media on the audience's psychological aspects have been conducted and researched more than a century ago. Giles (2003), in his work, cited Reeves and Anderson (1991), who explored ways in how psychological theory could illuminate media studies, and vice versa, arguing that ignoring the other (psychology and media) was difficult for each field. Media researchers could not overlook the cognitive processes involved in the use of media. For psychologists, cognitive and developmental psychology could be strengthened by considering media use, the same way reading studies have affected the general cognition theories (Giles, 2003).

6.2 What Is Media Psychology?

Media psychology uses psychology theories, principles, and techniques to research the effect mass media have on people, communities, and cultures (Fischoff, 2005). It is concerned with the interpersonal and intrapersonal psychological aspects that underlie the effect and use of any communication medium, regardless of the nature of the subject matter being communicated. In other words, it interacts with the psychological and social dimensions of interpersonal communication facilitated by some technology or catalyst. Dill (2013) defined it as the scientific study of human behaviour, thoughts, and feelings when using and developing the media. Media psychology, according to Giles (2003), is a specialty in the general psychological discipline. There are several types of psychology specialties: those that deal with specific aspects of human life (cognition, the social realm, and development of life); psychological (educational, clinical, and industrial/organizational) practices; and those studying specific areas of human activity (cultural, political, and media).

This field is applying the theories and research designs from cognitive, social, and developmental psychology to media study in individuals' lives (Polkinghorne, 2013). According to Dill (2013), media psychology uses psychology tools and theories to study and understand the complex interaction between human experience and media technologies, globally and individually. This involves applying theories and research of psychology, including methods and measurements, to combine them with theories and research of communication and media to understand individual differences and processes involved in media uses and their effects on individuals' cognitions, impacts, and behaviours. Rutledge (2010) stated that media psychology is considerably complex because it explores the incorporation of media technology into people's lives in various ways. Nowadays, people interact with media in multiple means as users, producers, and providers of all kinds of information across many platforms by using texts and photographs, music, and video, both synchronously and asynchronously. A positive contribution to life and society can be made by identifying the interactive and dynamic human-media relationships as a crucial way to a more precise and beneficial understanding of human and the media experience.

6.3 What Media Do to People?

Media can be pronounced in easy terms, like a television show is fascinating to watch. A book is great to be read, the Internet is useful for getting information, or social media is fantastic and fun when we can use them to catch up with long-lost contacted friends and family. Psychologists look into media from a theoretical perspective by applying social cognitive theory to the media, which means people are proactively interested in their development and can exercise control over their emotions, perceptions, and behaviour. Mass media such as television, radio, and the Internet are the best entertainment and extremely informative. Meanwhile, social media keeps us up to date with the goings-on around the world. According to Mehraj et al. (2014), individuals' attitudes and habits can be transformed with the use of media. Mehraj et al. (2014) sum up the functions of media as:

(a) The media may provide people with news and information that they require.
(b) The media can educate the public.
(c) The media can support the effective functioning of a democracy. Media may be used to notify the public about the government's policies and programs and how they can benefit from these programs. This can help people express their feelings and state of mind, which may help the government make necessary policy or program changes.
(d) The media can keep people around the world entertained.
(e) The media can act as agents of developmental change.
(f) The media can bring people from all over the world closer to one another.
(g) The media can help in promoting commerce and industry through advertising.
(h) The media can contribute to a country's political and democratic process.
(i) The media can bring about positive change in society.

A considerable body of research proposes that presentations in the media, including narratives of fictional works, can influence individuals' beliefs and attitudes (Green & Dill, 2013). According to Potter (2012) in his book 'Media Effect', six types of media effects may be applied to individuals: cognition, belief, attitude, influence, physiology, and conduct. Below are the personal effects of media, with their explanations, according to Potter (2012).

Cognitive media effect. This occurs when media exposure affects the mental processes of an individual or the result of those processes. The most readily recordable cognitive effect is the retention of information from media messages, particularly from books, journals, television news, and websites. Through the memorization process, the human mind can absorb this information. Other than that, the human mind may also be able to group the messages found in media in various ways to create new meanings. Besides, it can also generalize beyond media messages to generate real-life principles. All those mental activities are individual cognitive effects.

Beliefs. According to Potter (2012), Fishbein and Ajzen (1975) have defined faith as cognitions of the probability of association of an object or event with a given attribute. Faith is the assumption that something is right or is real. The media keeps on creating and shaping our faiths and beliefs by showing us the happening worldwide than we can see for ourselves directly. We all have assumptions about the reality of many things we have never seen now with our own eyes in our real lives, and many of these beliefs have come from media messages.

Attitudes. Attitudes are somewhat judgmental. People see a character in a film, for example, and make judgments about the attractiveness, heroic status, and likeability of that character. The need to make judgments on controversial/political issues or criticizing the products advertised in media and others is often caused when the media has presented narratives, news, or stories about real-world people, events, problems, and products.

Affect. It is referring to the feelings experienced by people, emotions and moods included. In particular, fear, lust, anger, and laughter can trigger emotions by the media. The media offers an opportunity for people to manage their moods. For example, by listening to music, we can relax, and by watching television, we may forget our struggles or difficulties for a while, even for a short time limit. If we feel stressed with all of our real lives problems, we can lose ourselves by playing video games or connecting to social media.

Physiological effect. It is an instantaneous or automatic bodily reflex or response. The response of the body can either be purely automatic or quasi-automatic. Automatic responses include dilation of the pupil, blood pressure, galvanic skin response, while quasi-automatic responses will consist of our heart rates or sexual responses to some stimuli. When we watch adventure/action/horror/thriller movies, for example, our heart rate and blood pressure will generally intensify, our muscles will automatically be feeling tense, and our palms will sweat. That is called a fight-or-flight reflex/response embedded in the brains of humans. Threats will cause consciousness as they trigger attention, and lastly, the body will prepare itself to fight or flee. This fight-or-flight reflex has permitted the human race to survive for thousands of years.

Behaviours. Generally, behaviours are characterized as an individual's overt acts (Albarracín et al., 2005). Researchers on media impacts have done several studies where they analysed media exposure habits of people to see which media they are using and how they use those media. Some studies expose individuals to specific media messages and track their subsequent actions for topics such as violence, the use of advertising goods, and political issues.

The study and research of media effects within the psychology and communication fields have a long history (Bartholow & Bolls, 2013; Bryant & Thompson, 2002). Lang (2009) conceptualized the use of media as a complex interplay between the human mind and the media's content. This conceptualization of media use has shifted the study of its effects onward by focusing on static media effects, incorporated in neural activity, on mental processes, which affect individuals with the impact of media content (Bartholow & Bolls, 2013).

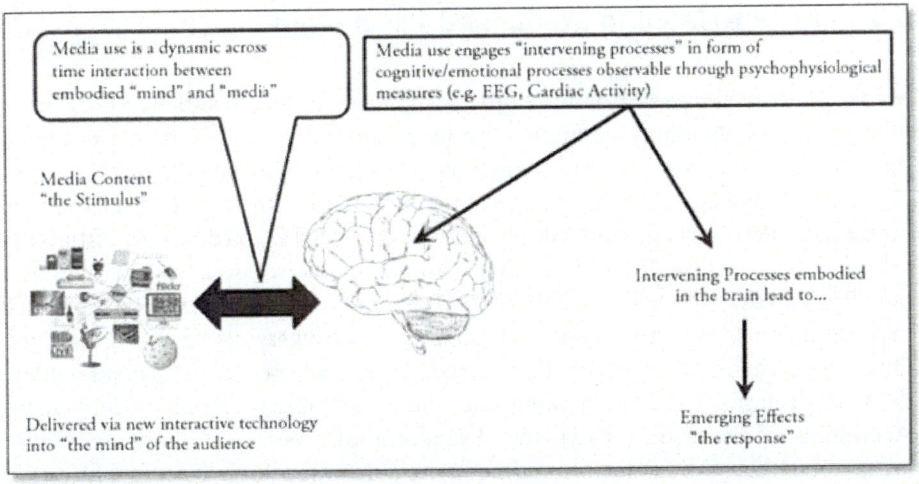

Fig. 6.1 Model depicting the stimulus-intervening processes-response approach to media effects (*Source* Bartholow and Bolls [2013])

Bolls et al. (2012) and Potter and Bolls (2012) argued that the psychophysiological perspective is essential to this theoretical standpoint of media use. It helped researchers monitor the mind's behaviour that has been replicated in the brain over time by using media, which provides critical data for the construction of novel and emerging interpretations of media effects. Figure 6.1 shows the model depicting the stimulus-intervening processes-response approach to media effects. This approach uses psychophysiology measures to examine the brain's activity, which functions as interfering processes between media content exposure and provable media effects.

Bartholow and Bolls (2013) argue that media and psychological studies are two distinctive disciplines that have made unique and exclusive contributions to research, such as how the brain processes media. Psychological scientists had been engaging in conducting a behavioural study involving how the influence of the environment and individual differences may affect the different ways the brain causes behaviour. The scholars in media studies are mainly keen to provide perspective into a wide range of mental processes involving media use in traditional media such as television and radio and new, interactive digital media forms (Bartholow & Bolls, 2013). In such a way, media studies and social psychology share a similar direction to comprehend how media content exposure can cause behavioural changes through its effects on internal states, such as cognition and emotions (Carnagey & Anderson, 2003).

6.4 Issues Related to Media and Psychology

Media plays a big part in how people respond to different situations, especially with the new technology additions. Media experiences are different according to the individual, culture, context, and trying to achieve (Boring-Bray, 2018). Some of the good impacts of media are that it enables people to understand the elements presented, improving communication among many people and different cultures. For example, by watching/listening to international news, we are kept informed about breaking news worldwide. Some shows and channels also offer educational programs that can increase our knowledge and make us more aware of the world around us. Other than television and radio, the Internet, and more specifically, social media, is the most prominent and commonplace to find a tremendous amount of information. According to Anderson (2018), social media allows people to discover topics or areas of interest while at the same time being able to reach, connect and meet other people. It has made communication much more accessible by providing the means to target all types of people's demographics. Because social media is easy to use, it can also be used to spread information rapidly by many users to reach more people.

Besides that, Boring-Bray (2018) also argued that people are getting smarter these days thanks to assisted learning and interactive media elements, particularly in video games. A study conducted at University College London and the Queen Mary University of London found that the result shows that the video game led participants to manage more information sources during gameplay, thus improving cognitive flexibility (Queen Mary University of London, 2013). Another study found that the improved learning of the regularity and environmental structure could act as a fundamental mechanism through which action video game affects efficiency in perception, attention, and cognition (Griffiths, 2014). Playing video games, in other words, enhances several skills, including paying attention.

Like anything good, however, there are some bad parts of it too. Our attention spans are decreasing because of the exposure to so much and fast-paced media. A study conducted by Microsoft in 2015 shows that the average amount of human attention span is found to have decreased from 12 s in 2000 to 8 s in 2013. This phenomenon is called 'The Goldfish Effect,' as a goldfish is thought to be able to keep its attention for 9 s. The research conducted in Canada in 2000 people and the brain activities of 112 others are studied by using EEG scans. Below is an excerpt from the report (Microsoft, 2015):

> While digital lifestyles decrease sustained attention overall, it is only true in the long-term. Early adopters and heavy social media users' front load their attention and have more intermittent bursts of great attention. They are better at identifying what they want/do not want to engage with and need less to process and commit things to memory.

According to Valentine (2018), there is a theorization of the mechanism underlying a causal interaction between 'media multitasking' and attention deficit. Susan Greenfield, an Oxford professor of synaptic pharmacology, argues that *social media risks 'infantilizing' the human mind* (Nusca, 2009). She also claimed that a sense of identity could be worn out by *"fast-paced, instant screen reactions, perhaps the next generation will define themselves by the responses of others.*" In a world of rapid and intermittent action, the brain must adapt to function over short periods. When socializing, people seem to have one eye (or both) on their phone; this shows that social norms are compromised, as people are increasingly distracted. According to a survey by The National Centre for Biotechnology Information and the National Library of Medicine in the USA, 79% of respondents used portable devices while watching television (known as dual-screening), and 52% checked their phones every half hour (Hooton, 2015). This constant bombardment of information and stimulation changes the nature of attention and how we interact in the digital world and real life.

Another issue to be discussed is that sleep time has decreased due to the use of media. According to the National Sleep Foundation of USA (2020), using televisions, smartphones, laptops, or other electronic devices before bed, may delay our body's internal clock (also known as circadian rhythm), inhibits the release of the sleep-inducing hormone melatonin, and therefore makes it harder to fall asleep. This is mostly to do with the short-wavelength, artificial blue light these devices emit. Those effects can lead to a significant, chronic sleep deficiency in the end. A study conducted by the Ontario Student Drug Use and Health Survey found that the more social media we use, the greater our likelihood are of a shorter duration of sleep (De Graaf, 2018). The survey results revealed that using social media for only one hour a day would seriously affect the amount of sleep the users had at night. According to Burke (2018), the sleep problem is caused by the blue light emission and additional psychological aspects of the use of social media to make a significant contribution to sleep deprivation. Social media networks are full of constant updates, and we can be tempted to check the newsfeeds repeatedly, as there will always be something new and exciting to look at. According to the National Health Service of the United Kingdom (2018), chronic sleep debt may result in long-term mood disorders such as depression and anxiety. Without having enough sleep, the brain and body systems will not function normally. Besides, sleep deprivation will also damagingly affect mental abilities and emotional state, and it may compromise decision-making processes and creativity (Watson & Cherney, 2019).

The next issue that should be considered is the addiction to media, particularly the Internet and social media. Internet addiction is a behavioural addiction where a person becomes dependent on using the Internet or other online devices, has an obsessive desire to spend much time on the Internet to the extent other areas of life such as relationships, work, or health suffer. Meanwhile, social media addiction can be regarded as an Internet addiction where people are compelled to excessively use social media (Griffiths, 2000; Starcevic, 2013). Meanwhile, individuals addicted to social media are usually too preoccupied with social media and fueled by an irresistible desire to log in and use social media (Andreassen & Pallesen, 2014). Hartney (2019) stated that addiction to the Internet is becoming universally recognizable and identified, especially in countries where it affects a significant number of people, such as South Korea, which has declared Internet addiction a national health problem. Much research has been carried out in Asia on Internet addiction and is a growing concern in developed countries in North America and Europe (Hartney, 2019). According to Hou et al. (2019), numerous studies on social media and mental health have revealed that prolonged use of social media is positively related to mental health problems (such as stress, anxiety, and depression) and negatively linked to long-term well-being. A study by Pantic et al. (2012) found that the time spent on social media has been linked significantly to symptoms of depression among high school students.

Other than that, one of the media-assisted crimes, identity theft, is also one of the crucial issues related to media use these days. Identity thieves are keen on bank account numbers, credit card numbers, medical records, passwords, and any other bit of information, which may allow our identity to be accessed. If they get the information they have been aiming for, identity thieves can ruin a person's credit rating, drain their bank accounts, and hold other personal/private information (Kagan, 2019). Identity thieves may use computer technology to attain other people's personal information. Spector (2018) stated that to find such information. They may search hard drives of stolen or discarded computers, hack computers or computer networks, and access computer-based public records. They use malware-gathering information to infect computers, search social media networking sites, or use misleading emails or text messages. The 2018 Identity Fraud Study released by Javelin Strategy and Research proved that 16.7 million people in the USA were victims of identity theft in 2017, up from 15.4 million in 2017, with $16.8 billion total stolen (Spector, 2018).

Next, exposure to media violence, including television, film, music, video games, and social media, poses a severe threat to users ' safety, particularly children and adolescents. Almost every single day, we are exposed to the abundance of violence, negativity, intolerance, and hatred in one form or another on the major news networks and social media. In recent years, social media can appear as if it has primarily become a central hub for spewing hate, intolerance, and in many cases, depicting 'real-life' acts of graphic violence and aggression indicative of a total disregard for human life (Pittaro, 2019). Media violence has led to people becoming desensitized to violent behaviour, expediting various sexual acts (Boring-Bray, 2018). According to the American Academy of Pediatrics (2009), thorough research evidence suggests that media violence can contribute to aggression, inappropriate behaviour, violent desensitization, nightmares, and fear of harm. Desensitization is the continuing decrease in response to an arousal stimulus as a constant exposure component (Krahé et al., 2011). In particular, desensitization to violent stimuli in the media is believed to reduce anxiety. Like other emotional responses, continued exposure to media violence and aggression can reduce adverse effects. As violent stimuli lose their ability to generate strong emotions, the stimulus is often portrayed (Anderson & Dill, 2000). Krahé et al. (2011) found that the more time people spent watching violent media portrayals, the less emotionally sensitive they will become too violent stimuli and the reduced amount of sympathy they have shown for violence victims (Mullin & Linz, 1995). According to Mrug et al. (2016), repeated exposure to violence can intensify violent behaviour by increasing aggressive behaviour. While exposure to media violence is not the only contributing reason to people's aggression and violence, it is a significant risk factor for psychological health that we need to think about and intervene.

6.5 What Can We Do?

We can make several efforts to ensure the well-being of our society in the context of media psychology aspects. They are listed as below;

(a) The media itself serves as the best control mechanism. We can use media to promote and circulate information on how media may or may not affect our psychological health.
(b) Awareness and prevention campaign. This involves activities and actions that can improve societal, psychological well-being. These actions include early childhood interventions, support for children, social assistance and support to the elderly populations; programs targeted at vulnerable people, mental health promotional activities in schools or work, violence prevention programs, and anti-discrimination laws and campaigns.
(c) Parental control for young users. Parental controls are meant to help protect children from inappropriate or harmful materials. These controls may be used in various ways, for instance, to ensure that children can only gain access to age-appropriate content, set usage times, and monitor and track their media use activities.
(d) Locus control to regulate self-control. Locus of control is the notion that can change a circumstance through behaviour. If individuals acquire the awareness, principles, ethics, and action skills needed, they will not partake in any dangerous or unethical acts by making improvements through their actions and behaviour.
(e) Digital Resilience. It is a crucial skill when people can think of strategies to help themselves if they are having a difficult time online, consider the impact they may have, and invent ways to build digital resilience. There are three areas of digital resilience, according to Parent Zone (2020). The first one is social and emotional skills and interpretation of the value of empathy and compassion in an offline and online world and what it means. Next is a safe space by being able to recognize and comprehend when at-risk online. Lastly is recovery, by understanding how to recover when things went wrong online (by recording and blocking) and learning from experiences.
(f) Designing Smarter Tech Tools Technology offers massive possibilities for learning, managing people's health, and connecting with others. New technology in educational gaming, health-care communication, robotics, and more is benefiting from psychologists' input (Weir, 2018).

6.6 The Future of Media Psychology

According to Eric Teller the CEO of Google's X research (Friedman, 2016), scientific and technological advances grew so slowly a thousand years ago that it could take a hundred years for the world to feel and look significantly different. However, by 1900, this process shifted and began to accelerate, and it only took 20–30 years for the technology to take one-step large enough to make the world unpleasantly different. He said,

> That time window-having continued to shrink as each technology stood on the shoulders of past technologies-has become so short that it is on the order of five to seven years from the time something is introduced to being ubiquitous and the world uncomfortably changing.

This concept applied to the development of media we see today, as there are always new media emerging from time to time in a short period, and luckily, we can adapt to the fast changes.

According to Luskin (2012), a licensed psychotherapist, media psychology's future is very bright. The conjunction of media, technology, communication, art, and science is transforming our world. Luskin (2012) proposed that there are several ways on how the potential media psychology in the future is best demonstrated by pointing out some of the current developments in the field. The first one is the improved incorporation of media psychology theories and courses into university graduate programs. Besides, substantial innovations in online-based education, which concentrate on understanding media psychology values and principles, are also crucial. We spent hours a day with screen media, yet we still know very little about how this screen time affects us. Thus, it is much recommended to add more knowledge in the psychological study of the causes and consequences of humans' media use.

There is also a need for research and advancement in telemedicine, telehealth, and tele therapy programs based on fundamental theories in media psychology. According to Suler (2002), face-to-face contact may be too daunting and stressful for some people. Text-based synchronous and asynchronous communication can focus more on their internal responses while still participating in therapeutic dialogue. These suggested techniques will permit people to feel less uncomfortable and insecure and have easier access to their issues. Electronic interaction promotes a sense of anonymity and confidentiality (Suler, 2002). This feature could attract people who do not feel confident in expressing emotions around someone, especially for such people who fear opening up to a new person and who typically would not attend therapy (Colbow, 2013).

Next, recognition of media psychology's importance in media programs, film, and television is also needed. By using the media as a medium to channel information relating to media psychology, we can ensure that the public will be knowledgeable in this area and thus be aware of their media utilization while taking care of their psychological health. Throughout the world, people must recognize how media, psychology, and behaviour are connected and how important they are to nearly every form of modern communication. By promoting a broad understanding of such elaborate social, health, family, and public issues, applying theories and methods in media psychology is necessary to improve public awareness of the importance of social and public health needs.

6.7 Conclusion

There are many benefits of media technologies towards our psychological aspects in life, but we are also becoming more aware of their risks and shortfalls. Rutledge (2010) stated that if we are to exercise media psychology, we ought to understand how media technologies and processes work; how they are created, generated, and utilized. Furthermore, we also need to understand psychology to start applying it to usability, efficacy, and impact issues. Media psychology uses psychological concepts and theories to analyse and understand the complicated relationship between human experience and media technologies. Thus, it is wise to understand how media and communication technologies may impact our lives' psychological aspects, maintain the positive impacts, and minimize (or eliminate) the adverse effects they will bring. To conclude this paper, we quote an anonymous technology developer/administrator who presented his/her view when being asked about Hopes for the Future of Digital Life in a research conducted by Pew Research Center (2018), who stated that,

> The harms brought by technology are considerable and should not be minimized. They represent both the adjustments that we need to make to accommodate new ways of doing things and structural changes and shifts in power. However, the benefits should not be forgotten; for every person who risks internet addiction or smartphone overload, some people see quantifiable advancements in the well-being and life quality, possibilities, education and human rights as a consequence of technology.

References

Albarracín, D., Zanna, M. P., Johnson, B. T., & Kumkale, G. T. (2005). Attitudes: Introduction and scope. In *The handbook of attitudes* (pp. 3–19). Lawrence Erlbaum Associates Publishers.

American Academy of Pediatrics. (2009). Media violence. *Council on Communications and Media, 124*(5), 1495–1503.

Anderson, B. (2018). *Social media communication.* The Medium. https://medium.com/the-benefits-of-communication/social-media-communication-df695b86f166.

Anderson, C. A., & Dill, K. E. (2000). Video games and aggressive thoughts, feelings, and behaviour in the laboratory and in life. *Journal of Personality and Social Psychology, 78,* 772–790.

Andreassen, C. S., & Pallesen, S. (2014). Social network site addiction—An overview. *Current Pharmaceutical Design, 20,* 4053–4061.

Baranowski, A., & Hecht, H. (2017). One hundred years of photoplay: Hugo Münsterberg's lasting contribution to cognitive movie psychology. *Projections, 11*(2), 1–21.

Bartholow, B. D. & Bolls, P. (2013). Media psychophysiology: The brain and beyond. In *The Oxford handbook of media psychology.* https://doi.org/10.1093/oxfordhb/9780195398809.013.0001.

Bolls, P. D., Wise, K., & Bradley, S. D. (2012). Embodied motivated cognition: A theoretical framework for studying dynamic mental processes underlying advertising exposure. In *Advertising theory* (pp. 105–119). New York: Routledge.

Boring-Bray, W. (2018). *What is the impact of media psychology?* https://www.betterhelp.com/advice/psychologists/what-is-the-impact-of-media-psychology/.

Bryant, J., & Thompson, S. (2002). *Fundamentals of media effects*. New York: McGraw-Hill.

Burke, C. (2018, January 26). Does social media affect your sleep? Science says the answer is definitely yes. *Elite Daily*. https://www.elitedaily.com/p/does-social-media-affect-your-sleep-science-says-the-answer-is-definitely-yes-8015656.

Carnagey, N. L., & Anderson, C. A. (2003). Theory in the study of media violence: The general aggression model. In D. Gentile (Ed.), *Media violence and children*. Westport, CT: Praeger.

Colbow, A. (2013). Looking to the future: Integrating telemental health therapy into psychologist training. *Training and Education in Professional Psychology, 7*(155).

De Graaf, M. (2018, January 24). Just one hour of social media a day is enough to ruin your sleeping pattern, study warns. *Daily Mail*. https://www.dailymail.co.uk/health/article-5304209/Just-ONE-HOUR-social-media-ruins-sleep.html.

Dill, K. E. (2013). Introduction. In *The Oxford handbook of media psychology*. https://doi.org/10.1093/oxfordhb/9780195398809.013.0001.

Fischoff, S. (2005). *Media psychology: A personal essay in definition and purview*. https://www.apadivisions.org/division-46/about/fischoff-media-psychology.pdf.

Fishbein, M., & Ajzen, I. (1975). *Belief, attitude, intention, and behaviour: An introduction to theory and research*. Reading, MA: Addison-Wesley.

Friedman, T. L. (2016). *Thank you for being late*. UK: Penguin Books.

Giles, D. (2003). *Media psychology*. New Jersey: Lawrence Erlbaum Associates Inc.

Green, M. C., & Dill, K. E. (2013). Engaging with stories and characters: Learning, persuasion, and transportation into narrative worlds. *The Oxford Handbook of Media Psychology*. https://doi.org/10.1093/oxfordhb/9780195398809.013.0001.

Griffiths, M. D. (2000). Internet addiction: Time to be taken seriously? *Addiction Research, 8*, 413–418.

Griffiths, M. (2014). Playing video games is good for your brain—here's how. *The Conversation*. https://theconversation.com/playing-video-games-is-good-for-your-brain-heres-how-34034.

Hartney, E. (2019). *How to know if you have an internet addiction and what to do about it*. Very Well Mind. https://www.verywellmind.com/internet-addiction-4157289.

Hooton, H. (2015, May 13). Our attention span is now less than that of a goldfish, Microsoft study finds. *The Independent*. https://www.independent.co.uk/news/science/our-attention-span-is-now-less-than-that-of-a-goldfish-microsoft-study-finds-10247553.html.

Hou, Y., Xiong, D., Jiang, T., Song, L., & Wang, Q. (2019). Social media addiction: Its impact, mediation, and intervention. *Cyberpsychology: Journal of Psychosocial Research on Cyberspace, 13*(1).

Kagan, J. (2019). *Identity theft*. Investopedia. https://www.investopedia.com/terms/i/identitytheft.asp.

Krahé, B., Möller, I., Huesmann, L. R., Kirwil, L., Felber, J., & Berger, A. (2011). Desensitization to media violence: Links with habitual media violence exposure, aggressive cognitions, and aggressive behaviour. *Journal of Personality and Social Psychology, 100*(4), 630–646.

Lang, A. (2009). The limited capacity model of motivated mediated message processing. In R. L. Nabi & M. B. Oliver (Eds.), *The Sage handbook of media processes and effects* (pp. 193–204). Thousand Oaks, CA: Sage.

Luskin, B. J. (2012). *E = Enhanced Media: The psychology of eLearning through enhanced media*. Luskin's Learning Psychology Series—No. 3. https://www.psychologytoday.com/us/blog/the-media-psychology-effect/201207/eenhanced-media.

Mehraj, H. K., Bhat, A. N., & Mehraj, H. R. (2014). Impacts of media on society: A sociological perspective. *International Journal of Humanities and Social Science Invention, 3*(6), 56–64.

Microsoft. (2015). *Attention spans*. Consumer Insights, Microsoft Canada. Retrieved from http://dl.motamem.org/microsoft-attention-spans-research-report.pdf.

Mrug, S., Madan, A., & Windle, M. (2016). Emotional desensitization to violence contributes to adolescents' violent behaviour. *Journal of Abnormal Child Psychology, 44*(1), 75–86.

Mullin, C. R., & Linz, D. (1995). Desensitization and resensitization to violence against women: Effects of exposure to sexually violent films on judgments of domestic violence victims. *Journal of Personality and Social Psychology, 69*, 449–459.

Münsterberg, H. (1916). *The photoplay: A psychological study*. New York: Appleton.

National Health Service. (2018). *Why lack of sleep is bad for your health*. https://www.nhs.uk/live-well/sleep-and-tiredness/why-lack-of-sleep-is-bad-for-your-health/.

National Sleep Foundation. (2020). *Why electronics may stimulate you before bed*. https://www.sleepfoundation.org/articles/why-electronics-may-stimulate-you-bed.

Nusca, A. (2009). *Will Facebook 'infantilize' the human mind?* CBS Interactive. https://www.zdnet.com/article/will-facebook-infantilize-the-human-mind/.

Pantic, I., Damjanovic, A., Todorovic, J., Topalovic, D., Bojovic-Jovic, D., Ristic, S., et al. (2012). Association between online social networking and depression in high school students: Behavioural physiology viewpoint. *Psychiatria Danubina, 24*, 90–93.

Parent Zone. (2020). *What is digital resilience and wellbeing? Parent zone's resilience and wellbeing*. https://parentzone.org.uk/article/what-digital-resilience-and-wellbeing.

Pew Research Center. (2018). *The future of well-being in a tech-saturated world*.

Pittaro, M. (2019, May). Exposure to media violence and emotional desensitization. What are the long-term consequences to children and adolescents? *Psychology Today*. https://www.psychologytoday.com/us/blog/the-crime-and-justice-doctor/201905/exposure-media-violence-and-emotional-desensitization.

Polkinghorne, D. E. (2013). Qualitative research and media psychology. *The Oxford Handbook of Media Psychology*. https://doi.org/10.1093/oxfordhb/9780195398809.013.0001.

Potter, W. J. (2012). *Media effects*. USA: Sage.

Potter, R. F., & Bolls, P. D. (2012). *Psychophysiological measurement and meaning: Cognitive and emotional processing of media*. New York: Routledge.

Queen Mary University of London. (2013). *Playing video games can boost brain power*. American Association for the Advancement of Science (AAAS). https://www.eurekalert.org/pub_releases/.

Reeves, B., & Anderson, D. R. (1991). Media studies and psychology. *Communication Research, 18*, 597–600.

Rutledge, P. (2010). *What is media psychology? And why you should care*. Media Psychology Research Center. https://www.apadivisions.org/division-46/about/rutledge-media-psychology.pdf.

Spector, N. (2018). *What is identity theft? Protection, how to report fraud, and more*. NBC News. https://www.nbcnews.com/storyline/smart-facts/.

Starcevic, V. (2013). Is Internet addiction a useful concept? *Australian and New Zealand Journal of Psychiatry, 47*, 16–19.

Suler, J. R. (2002). Identity management in cyberspace. *Journal of Applied Psychoanalytic Studies, 4*, 455–459.

Valentine, S. (2018). *The goldfish effect: Why social media shortens our attention span*. Educational Advice Educational News MyTutor for Parents. https://www.mytutor.co.uk/blog/attention-span-social-media/.

Watson, S., & Cherney, K. (2019). *The effects of sleep deprivation on your body*. Healthline. https://www.healthline.com/health/sleep-deprivation/effects-on-body#1.

Weir, K. (2018). Designing smarter tech tools. *Monitor on Psychology, 49*(10), 72.

Dr. Huzili Hussin is an Associate Professor and Head of Centre of Excellence for Social Innovation and Sustainability at Faculty of Applied and Human Sciences, Universiti Malaysia Perlis. His area of expertise is counselling psychology especially in cognitive-behavioural therapy Intervention. He also the panel of reviewers for national and international journals such as *Journal on Leadership and Policy* from Dr. Mahathir Institute of Thought, UUM, Asia Pacific Research Journal, Far-East South-east Asia Center, University of Sindh Pakistan and International Research Journal of Arts and Humanities, Faculty of Arts, University of Sindh Pakistan. Besides, he has been appointed to the Malaysian Qualification Agency's Accreditation (MQA) as a panel reviewer in the area of counselling psychology. Based on this expertise, he has been appointed as committee members of (MJA) Accreditation Councils Committee by MQA and a Foreign External Peer Reviewer at Khulna University of Engineering in Bangladesh. Dr. Huzili has also been appointed as Adjunct Professor at the University of Ubudiah Indonesia, Acheh Indonesia and Visiting Scholar at the University of Sindh in Pakistan. He is also a visiting lecturer of Universiti Malaysia Sabah for the Master of Psychology Counselling program. He was also involved in community services for the country. Dr. Huzili has been appointed as an Advisor for the Juvenile Court, a member of Board Visitors for the Malaysian Prison Department, and a panel for deradicalisation by the Minister of Home Affairs. Currently, he is a consultant for the Counselling and Psychology Division, Perlis State Social Welfare Department and columnist for Sinar Harian newspaper every Saturday.

Adila Ismail is a lecturer at the Faculty of Applied and Human Sciences, Universiti Malaysia Perlis. Her expertise is Communication and specialised in social media psychology. She has published a few journal articles that were indexed by SCOPUS. She was appointed as program chairman for the New Media Communication program for 4 years and currently she is pursuing her Ph.D. in media psychology in Universiti Malaysia Perlis.

Dr. Mohammad Rezal Hamzah is a senior lecturer and Head Cluster for New Media Communication and Technology at Centre of Excellence for Social Innovation and Sustainability, Faculty of Applied and Human Sciences, Universiti Malaysia Perlis. Currently he is a Deputy Dean for Industrial Collaboration and Quality Management and he was a Deputy Dean for Academic and Research from 2017 until 2020. His expertise is communication and specialised on health communication and new media for the young generation. He has published several books and chapters as well as articles journals that were indexed by WOS and SCOPUS. He has been appointed as a reviewer for several journals and also academic advisor for the program of Public Relation and Media Contemporary Universiti Sultan Zainal Abidin, Terengganu.

Chapter 7
Critical Success Factors of Adopting Servitization Strategy

Anas Hasbullah and Norani Nordin

Abstract Currently, the industry focuses on technology and innovation to increase the capabilities and relevance to markets. However, there is an important strategy that needs attention, especially in the manufacturing sector, to improve the company's performance. "Servitization" is a strategy to add value proposition to a product through service offerings. This chapter explores the concept of servitization strategy and identifies important factors to adopt the strategy. The concept of product and service is explained in this chapter, and the evolution of research related to this topic. Six critical success factors of servitization identify as the outcome of this study.

7.1 Introduction

In striving towards a high-income nation status by 2020, the Malaysian manufacturing sector remains a core sector for sustainable growth under the 11th Malaysia Plan. The manufacturing industry contributed to 25% of the country's gross domestic product (GDP) while the services industry contributes 57% (Department of Statistics Malaysia, 2020). However, the Malaysian manufacturing industry sectors are positively expanding despite facing many challenges and under massive pressure. According to the Ministry of Finance Malaysia, in the Economic Outlook 2020 Report, efforts to boost the high-value and smart manufacturing industry are being promoted seriously by the government.

Accordingly, this chapter will elaborate on "servitization" as a mechanism to strengthen the manufacturing industries in Malaysia. Servitization is a strategy that transitions a manufacturing company into becoming a service-oriented company or adopting a service business model where the company gradually offers services in

A. Hasbullah (✉) · N. Nordin
School of Technology Management and Logistics, Universiti Utara Malaysia, Changlun, Malaysia
e-mail: anas.upm10@gmail.com

N. Nordin
e-mail: rani@uum.edu.my

N. F. Elias et al. (eds.), *Service Excellence for Sustainability*,
https://doi.org/10.1007/978-981-16-2579-4_7

addition to its existing products. This strategy is important since currently, manufacturers are having difficulties to survive by relying only on the selling of products. More technologically advanced countries have actively promoted and implemented this strategy decades ago. On top of that, they have successfully transformed their business (manufacturing) into the service business. However, in developing nations servitization is a new concept that is just catching on.

7.2 Background of Servitization

Servitization is a concept where a company changes its strategy towards service-based to generate revenue streams. As service is dominating strategic thinking of business, service has been seen as the strategy for manufacturing survival for the developed economies such as Western and European countries. This is due to competition with emerging economies such as China which offers products at a lower price (Baines & Lightfoot, 2013). This factor drives manufacturing to move up the value chain, seeking to innovate and create more advanced products and services, so they do not compete based on cost alone. The driven factor of servitization started in the early 90th in western countries where manufacturers have to reduce their production costs to compete with low-income countries producers. This shows that developing economies such as China forced the developed economies such as the United States to compete based on product pricing. Hence this supports the factor why the manufacturing shift towards services. Thus, servitization expands the addition of services to manufacturers' core product offerings to create additional customer value (Raddats et al., 2019).

Besides, the rise of the services economy, the emergence in the dominance of the service sector of most developed economies, and the decline in manufacturing as a percentage of GDP drive many manufacturing companies to consider moving to service business (Young, 2012). According to Belal et al. (2014), pure manufacturing businesses find it hard to sell their products because of the increase of competencies from major competitors and developing economies. In short, to survive in the globalisation era, manufacturers and business owners extend their services offering to seek competitive advantage and gain profits for their companies and maintain their customer relationship. The growth of the service industry led many non-service companies, for example, manufacturers to transform into product-service systems (PSS) whereby they bundle the products with services.

7.3 Definition of Servitization

The term servitization coined by Vandermerwe and Rada (1988) is widely known as "creating value by adding services to the products". It is defined as a process of moving from traditional product or service focus to an "integrated product & service

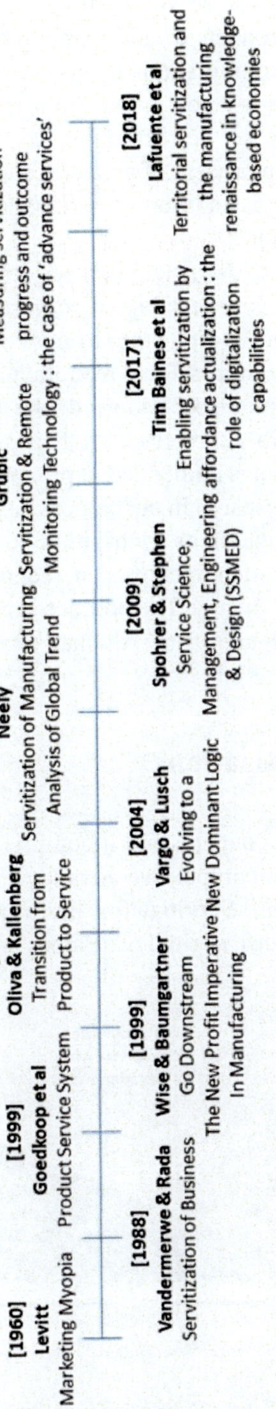

Fig. 7.1 Evolution of product service research

offering" that manufacturers offer for the customers. The concept can be defined as a combination of products and services where manufacturers offer "package" or "bundles" of services and products to add value to the products. These customer-focused bundles consist of combinations of goods, services, self-service, support, and knowledge. Servitization is a continuous improvement initiative by a company to offer better service and new service offerings. According to Salonen et al. (2017), servitization is not a transition process of a manufacturing company into becoming a service company, and rather it is an improvement or advancement of existing service provisions within the company.

Similarly, Baines and Shi (2015) state that servitization is not only focusing on manufacturing industries. It can be adopted in various industries, but it is more prevalent within manufacturers. Despite the strong prevalence of servitization among manufacturing firms, servitization affects manufacturing firm performance (Zhou et al., 2020). Servitization can also be viewed as a business process and strategy where business owners integrate services with their products. As such, servitization can be seen to be a concept that is embedded in product-service systems. This can be observed in the evolution of research in the field, as described in Fig. 7.1. It shows the growth of product-service research, its contribution to the manufacturing and service industry, how it changed the manufacturers' perception, and how the organisation transforms into a servitized company. The product-service is the key strategy to build service elements within an organisation (Ahamed et al., 2013).

7.4 Concepts of Servitization

Servitization is an intriguing concept, and it comprises various factors to form the backbone of a manufacturer's competitive advantage. Figure 7.2 shows the knowledge paradigm of servitization. Servitization lies at the confluence of the product worldview and the pure services worldview. Baines and Shi (2015) highlighted that

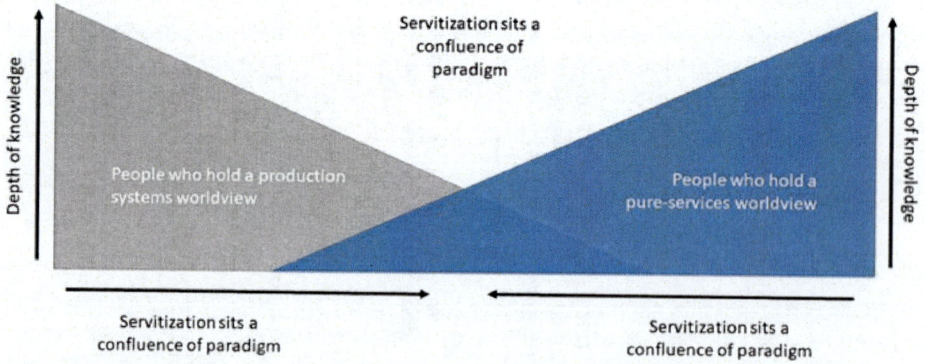

Fig. 7.2 Servitization paradigm (*Source* Baines and Lightfoot [2013])

incremental change in management and operations has to make way to a paradigm shift in working to attain servitization. It takes a lot of techniques and strategies for the company to be servitized.

As part of a product-service system, servitization can be viewed in a continuum ranging from services that support products to services that support customers. In such a scenario, companies offer three different categories of product-service propositions to their customers. The first category, known as the base service, caters to customers who want to do things themselves. Here, the company will focus on product provision, which includes product/equipment provision, spare parts provision, and warranty. The second category is the intermediate service, which serves customers who want to do things with the company. In such a situation, the company will focus on the maintenance of a product condition. Services offered include help desk services, training, maintenance, repair, and overhaul. The final category is more complex, known as the advanced service, which serves customers who want the company to do things for them (Baines & Lightfoot, 2013). Here, customers are concerned with the capability of a product that is supplied by the manufacturer for their use. Hence, manufacturing companies need to focus on the capability of the products. Examples of services in this category include customer support agreement, risk and reward sharing contract, and revenue-through-use contact. This continuum of product-service propositions is depicted in Fig. 7.3.

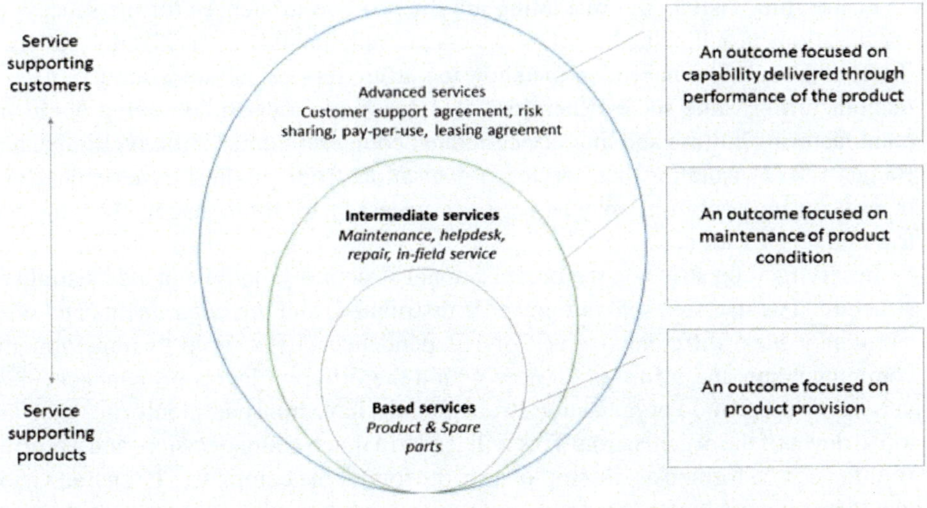

Fig. 7.3 The categories of product-service offered by manufacturer (*Source* Baines and Lightfoot [2013])

7.5 Critical Factors of Adopting Servitization Strategy

Servitization involves the process of better services offering, mutual customer understanding, increase loyalty, and ensure sustainability. This strategy may be cost-effective and may require a significant investment, but as a result, the company will gain higher profit and competencies (Benedettini et al., 2015). However, the risk is high, and failure may occur. In Malaysia, many manufacturing companies have service offerings but have limited knowledge and have difficulties extending their service. To identify the critical factors for adopting servitization strategy among companies, the authors identified 30 articles with most significant topics and issues on the servitization transition process published from the year 2000–2020. Table 7.1 shows the selected articles.

From the articles mentioned in Table 7.1, the keywords described in the process of servitization were extracted. The frequency of each keyword as it appeared in the articles is counted and subsequently clustered as "themes". Table 7.2 shows the list of the keywords with the respective frequencies.

The keywords shown in Table 7.2 are then further reviewed and filtered to get the most significant theme for the critical success factor of adopting the servitization strategy. Some of the keywords are identical and have the same meaning. Then, the authors combined the keywords under the same theme. Table 7.3 shows the final core themes identified from the analysis.

Leadership, vision, and marketing are the positive influences for the success of organisations in achieving the effectiveness of implementing a servitization strategy. Some of the scholars have identified how to nurture service culture without replacing manufacturing value set. Furthermore, this can avoid a clash between a dominant manufacturing culture and the service-related counterculture. Similarly, Bigdeli and Baines (2017) isolating the service organisation from product development and manufacturing operations may be a critical success factor for managing the transition towards servitization.

In servitization strategy, the organisational structure should be in a decentralised structure. The decision making power is distributed, and the departments and divisions may have different degrees of independence. To be able to transform the communication and information flow within the different levels of management is very important. Good organisation structure must have transparent information from top to down of the organisation. This will lead to a much better decision, and a holistic solution can be formed as the employee is the root of the company. The shared vision would be effectively delivered to all employees and they will strive towards the same thinking and objectives.

Employees are the vital resources of a company. Workplace communication is essential to companies because it allows companies to be productive and operate effectively. Employees can experience an increase in morale, productivity, and commitment if they can communicate up and down the organisation's communication chain. This consists of oral, written, and non-verbal. To ensure the sustainability of a company, the key employee must be well treated. The employer should

Table 7.1 Selected articles on servitization transition process

No	Author	Article Title	Sources	Year
1	Andrew Davies	Moving base into high-value integrated solutions: a value stream approach	Industrial and Corporate Change	2004
2	Charlotte Whndahl	Manufacturing firms and integrated solutions: characteristics and implications	*European Journal of Innovation Management*	2004
3		Developing integrated solutions: a network perspective	Proceedings of the 21st IMP Conference	2005
4	Andy Neely	The servitization of manufacturing: an analysis of global trends	Centre for Business Performance	2007
5		Exploring the financial consequences of the servitization of manufacturing exploring the financial consequences of the	*Operation Management Research*	2008
6.		The servitization of manufacturing: Further evidence	18th European Operations Management Association Conference, Cambridge	2011
7.	Tim Baines	State of the art of product service system	*Proceedings of the Institution of Mechanical Engineers, Part B: Journal of Engineering Manufacture*	2007
8.		Towards operation strategy for product centric servitization	*International Journal of Operations & Production Management*	2009
9.		The servitization of manufacturing: a review of literature and reflection on future challenges	*Journal of Manufacturing Technology Management*	2009
10.	Tim Baines	Servitization within manufacturing: exploring the provision of advanced services and their impact on vertical integration	*Journal of Manufacturing Technology Management*	2011
11		Servitization of manufacture: Exploring the deployment and skills of people critical to the delivery of advanced services	*Journal of Manufacturing Technology Management*	2012

(continued)

Table 7.1 (continued)

No	Author	Article Title	Sources	Year
12		Servitization impact study How UK based manufacturing organisations are transforming themselves to compete through advanced services	Aston-Servitization.Com	2013
13.		Servitization: revisiting the state-of-the-art and research priorities	*International Journal of Operations & Production Management*	2013
14.		Made to Serve. How manufacturers can compete through servitization and product-service systems	Statewide Agricultural Land Use Baseline 2015	2013
15.		The servitization of manufacturing: a systematic literature review of interdependent trends	*International Journal of Operations & Production Management*	2013
16.		A Delphi study to explore the adoption of servitization in UK companies	Production Planning & Control	2015
17.		Servitization: revisiting the state-of-the-art and research priorities	*International Journal of Operations and Production Management*	2017
18.	Veronica Martinez	Challenges in transforming manufacturing organisations into product-service providers	*Journal of Manufacturing Technology Management*	2010
19.		The future of servitization: technologies that will make a difference	Cambridge Service Alliance	2015
20.		Seven Critical success factors in the shift to services	Cambridge Service Alliance	2016
21.	Zahir Ahamed	A strategic journey of firm transformation towards a new framework for implementing servitization strategy	*International Journal for Infonomics*	2012
22.		The servitization of manufacturing: an empirical case study of IBM Corporation	*International Journal of Business Administration*	2013
23.		Organizational factors to the effectiveness of implementing servitization strategy	*Journal of Service Science and Management*	2013

(continued)

Table 7.1 (continued)

No	Author	Article Title	Sources	Year
24.	Fredrik Nordin	Critical issues during servitization: an in-depth case study	International Service Research Conference	2012
25.	H.M. Belal	Knowledge space concept and its application for servitizing manufacturing industry	Journal of Service Science and Management	2012
26.		Strategic corporate alliance framework for servitization	Knowledge Co-Creation	2012
27.		Value co-creation with customer through recursive approach based on Japanese Omotenashi service	*International Journal of Business Administration*	2013
28.		Infrastructure innovation to attain service value sustainability: viewpoint of resource management	*International Journal of Service Science, Management, Engineering, and Technology*	2014
29.	Ali Ziaee Bigedli	Holistic approach to evaluating servitization: a content, context, process framework	22nd EurOMA Conference—Operations Management for Sustainable Competitiveness	2015
30.		Organisational change towards servitization: a theoretical framework	*Competitiveness Review: An International Business Journal*	2017
31.	Marco Perona	Research vs. practice on manufacturing firms' servitization strategies: a gap analysis and research agenda	Systems	2017

take responsibility for employee well-being, health, emotion, motivation, education, skills, and experience. The employee should have the right opportunities to grow and learn to be a better person. This will be beneficial to the company as the performance will be improved. Employers should build an excellent platform for the employee to learn and work; the employer must show a good example of leadership and a close relationship with the employee.

Strategic direction is a course of action that leads to the achievement of the goals of an organisational strategy. The strategy determines the contents, approach, and the character of the organisation's activities. The strategy is a plan, the tactics will determine how the plan will be executed, and the vision is the result and must be suitable for the company's capabilities. A measurable step must be identified to achieve the objective and plan for the company.

In a servitization model, the value chain is a competitive strategy to attain partners' and customers' loyalty through multichannel collaboration. The company can also work together with their competitors by collaborating in product and service offerings that are of value and valued and beneficial to both parties. This paves the way to

Table 7.2 The frequency of keywords appeared in the articles by selected authors

Researchers / Keywords	Andy Neely	H.M. Belal	Marco Perona	Fredrik Nordin	Tim Baines	Veronica Martinez	Ahamed Zahir	Ali Ziaee Bigdeli	Andrew Davies	Charlotte Windahl	Frequency
Cultural shift	/			/	/	/	/	/	/	/	8
Customer centricity		/	/	/		/	/			/	6
Customer orientation			/	/		/	/		/	/	6
Customer value			/	/		/	/		/	/	6
Employee empowerment	/				/	/	/	/		/	6
Employee skillset	/	/			/	/	/	/		/	7
Human resources	/	/	/		/	/	/	/		/	8
Leadership			/	/	/		/	/		/	6
Mindset		/		/		/	/	/		/	6
Operation strategies		/	/	/	/	/			/		6
Organisation management				/	/	/	/		/	/	6
Organisational philosophy		/		/	/	/	/		/		6

(continued)

Table 7.2 (continued)

Researchers / Keywords	Andy Neely	H.M. Belal	Marco Perona	Fredrik Nordin	Tim Baines	Veronica Martinez	Ahamed Zahir	Ali Ziaee Bigdeli	Andrew Davies	Charlotte Windahl	Frequency
Organisational strategy	/	/	/	/	/	/	/			/	9
Organisational structure	/			/	/	/	/	/		/	7
Organisational transformation		/	/	/	/	/	/			/	7
Partners	/	/			/	/	/	/		/	7
Service culture	/	/		/	/	/	/	/	/	/	9
Service design structure	/			/	/	/	/	/	/	/	8
Service integration		/	/	/		/	/		/	/	7
Service process delivery	/		/	/		/	/			/	6
Strategic direction	/	/	/	/	/	/	/	/			8
Strategic partnership	/	/	/	/	/	/	/	/		/	9
Strategies	/	/	/		/	/	/				6
Supplier relationship		/		/	/	/	/	/		/	7

(continued)

Table 7.2 (continued)

Researchers / Keywords	Andy Neely	H.M. Belal	Marco Perona	Fredrik Nordin	Tim Baines	Veronica Martinez	Ahamed Zahir	Ali Ziaee Bigdeli	Andrew Davies	Charlotte Windahl	Frequency
Supply chain management		/			/	/	/	/		/	6
Top management support			/			/	/	/	/	/	6
Value added	/	/		/		/	/	/			6
Value chain	/	/		/	/	/	/	/	/	/	9
Vision	/	/		/	/	/	/	/		/	8

Table 7.3 Final core theme for critical success factor of adopting servitization strategy

Core theme	Critical success factor	Description
Vision/philosophy/mindset	Vision	Service focus
Organisation Structure/management/service culture	Organisation structure	Flat structure, project-based structure, team leader make decision, service culture, improve communication, information flow, knowledge sharing
Employee/culture/leadership	Employee management	More on service department, high-level skills
Value chain/partnership/supplier	Strategy	New services and product, service innovation, customer relationship management, total solution, improve efficiency, product service system
Strategy/ service integration	Value chain	Outsourcing, share value, profit and cost-sharing, value co-creation, co-production, customer centricity, impact community, environment-friendly, improve product lifecycle
Customer orientation/value added	Customer empowerment	Relationship-based, value co-creation, customer-centric, full customer support

create more allies than competitors. The aim is to create a mutual value agreement that benefits the participants in the value chain circle. The success key of the value chain is the collaboration and trust between the parties.

The customer is the person who pays for goods or services from a business. In this context, the customer usually refers to business customers and end-users or consumers. Customer-centric business is where the customer as a co-creator of the product or services. Feedback and response from the customers are gathered, and the business offerings can be designed based on input from customers. Direct involvement from the customers can lead to customisation of products or services based on the customers' precise specification or requirements. Business owners need to engage customers to get information about the products and services. If the customer is not satisfied, and the product or service does not fit the customer's needs or expectation, it would be of no value to the customer. Let the customer be the decision-maker of the product or service they wanted, by observing the product efficiency and the prices of the two firms, and customers can choose options to satisfy their needs (Wang et al., 2019).

7.6 Conclusion

In the business context, service is a tangible logical business process that creates value for the entity's benefits and other entities. Due to globalisation and shrinking profits from manufactured products, manufacturing companies must find different dimensions of competition. Service is seen as the strategic solution, and this leads them to adopt the concept of servitization strategy. In this chapter, the origins and the definition of servitization is discussed. Finally, based on a review of relevant literature, the critical success factors of adopting servitization strategy within organisations are identified and explained. In conclusion, there are six critical success factors of adopting servitization strategy: vision, organisation structure, employee management, strategy, value chain, and customer empowerment.

References

Ahamed, Z., Kamoshida, A., Inohara, T., Zahir, A., Akira, K., & Takehiro, I. (2013). Organizational factors to the effectiveness of implementing servitization strategy. *Journal of Service Science and Management, 6*(June), 177–185.

Baines, T. S., & Lightfoot, H. (2013). Made to Serve. How manufacturers can compete through servitization and product-service systems. In *Statewide agricultural land use baseline 2015* (Vol. 1). https://books.google.co.uk/books?id=F43UOJGSAGMC.

Baines, T. S., & Shi, V. G. (2015). A Delphi study to explore the adoption of servitization in UK companies. *Production Planning & Control, 7287*(August), 1171–1187. https://doi.org/10.1080/09537287.2015.1033490.

Belal, H. M., Shirahada, K., & Kosaka, M. (2014). Infrastructure innovation to attain service value sustainability: Viewpoint of resource management. *International Journal of Service Science, Management, Engineering, and Technology, 5*(June), 19–35.

Benedettini, O., Neely, A., & Swink, M. (2015). Why do servitized firms fail? A risk-based explanation. *International Journal of Operations & Production Management, 35*(6), 946–979.

Bigdeli, Z. A., & Baines, T. (2017, June). Organisational transformation towards servitization. In *Proceedings of the Spring Servitization Conference (SSC2017)*. Retrieved from https://www.researchgate.net/profile/Ali_Ziaee_Bigdeli/publication/317559945_Organisational_Transformation_towards_Servitization/links/593ff51ea6fdcc1b10c2cc0b/Organisational-Transformation-towards-Servitization.pdf [19 November 2020].

Department of Statistics Malaysia. (2020). *Malaysia economic performance third quarter 2020*. https://www.dosm.gov.my/v1/index.php [15 January 2021].

Raddats, C., Kowalkowski, C., Benedettini, O., Burton, J., & Gebauer, H. (2019). Servitization: A contemporary thematic review of four major research streams. *Industrial Marketing Management, 83*(April), 207–223. https://doi.org/10.1016/j.indmarman.2019.03.015.

Salonen, A., Saglam, O., & Hacklin, F. (2017). Servitization as reinforcement, not transformation. *Journal of Service Management, 28*(4), 662–686.

Vandermerwe, S., & Rada, J. (1988). Servitization of business: Adding value by adding services. *European Management Journal, 6*(4), 314–324.

Wang, K., Huang, T., & Jiang, Z. (2019). Competitiveness of servitization with new transaction base. *IFAC-PapersOnLine, 52*(13), 1216–1221. https://doi.org/10.1016/j.ifacol.2019.11.364.

Young, L. (2012). *From products to services: Insight and experience from companies which have embraced the service economy. In from product to service*. London: Wiley.

Zhou, D., Yan, T., Zhao, L., & Guo, J. (2020). Performance implications of servitization: Does a Manufacturer's service supply network matter? *International Journal of Production Economics, 219*(May 2019), 31–42. https://doi.org/10.1016/j.ijpe.2019.05.019.

Anas Hasbullah was a Master student of School of Technology Management and Logistics. His thesis was in "Servitization Strategy of Malaysia Manufacturing Industry". Currently, he is working as Research Assistant at Research and Innovation Management Research Centre (RIMC) in Universiti Utara Malaysia.

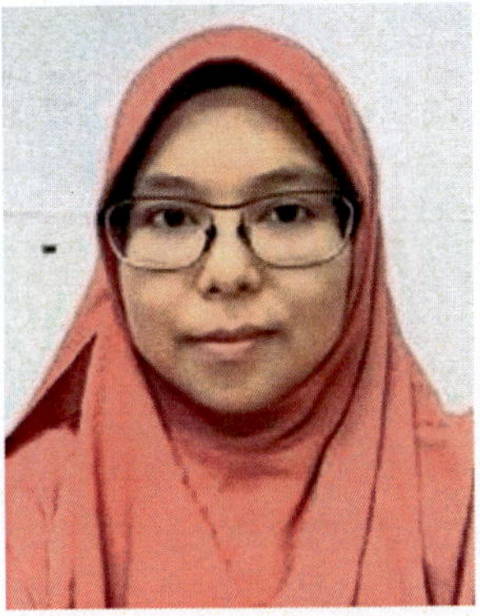

Dr. Norani Nordin is an Associate Professor of School of Technology Management and Logistics, Universiti Utara Malaysia. She is currently appointed as one of the UUM's AACSB Managers and the Leader of Knowledge Science Research Lab. Her works focus especially on Lean Management, Service Science and Knowledge Co-creation. Norani Nordin has published 80 articles in high impact journals, conference proceedings and book chapters.

Index

N. F. Elias et al. (eds.), *Service Excellence for Sustainability*,
https://doi.org/10.1007/978-981-16-2579-4

T

U

V

W